Lecture Notes in Statistics 76

Edited by J. Berger, S. Fienberg, J. Gani, K. Krickeberg, I. Olkin, and B. Singer

Lennart Bondesson

# Generalized Gamma Convolutions and Related Classes of Distributions and Densities



Springer-Verlag
New York Berlin Heidelberg London Paris
Tokyo Hong Kong Barcelona Budapest

Lennart Bondesson
Department of Forest Biometry
Swedish University of Agricultural Sciences
S-90183, Umeå, Sweden

AMS Subject Classifications (1980): 60E05, 60E07, 60E10, 62E10, 60G50, 60G55, 44A10

Library of Congress Cataloging-in-Publication Data
Bondesson, Lennart.
Generalized gamma convolutions and related classes of distributions and densities / Lennart Bondesson.
p. cm. — (Lecture notes in statistics ; v. 76)
Includes bibliographical references and indexes.
ISBN 0-387-97866-6
1. Convolutions (Mathematics) 2. Distribution (Probability theory) I. Title. II. Series: Lecture notes in statistics (Springer-Verlag) ; v. 76.
QA273.6.B66 1992
519.2'4—dc20 92-16277

Printed on acid-free paper.



Camera ready copy provided by the author.
Printed and bound by Edwards Brothers, Inc., Ann Arbor, MI.
Printed in the United States of America.

9 8 7 6 5 4 3 2 1

ISBN 0-387-97866-6 Springer-Verlag New York Berlin Heidelberg
ISBN 3-540-97866-6 Springer-Verlag Berlin Heidelberg New York

To OLOF THORIN

# PREFACE

Generalized Gamma convolutions were introduced by Olof Thorin in 1977 and were used by him to show that, in particular, the Lognormal distribution is infinitely divisible. After that a large number of papers rapidly appeared with new results in a somewhat random order. Many of the papers appeared in the Scandinavian Actuarial Journal. This work is an attempt to present the main results on this class of probability distributions and related classes in a rather logical order. The goal has been to be on a level that is not too advanced. However, since the field is rather technical, most readers will find difficult passages in the text. Those who do not want to visit a mysterious land situated between the land of probability theory and statistics and the land of classical analysis should not look at this work.

When some years ago I submitted a survey to a journal it was suggested by the editor, K. Krickeberg, that it should be expanded to a book. However, at that time I was rather reluctant to do so since there remained so many problems to be solved or to be solved in a smoother way than before. Moreover, there was at that time some lack of probabilistic interpretations and applications. Many of the problems are now solved but still it is felt that more applications than those presented in the work could be found. I hope that the readers will be able to find many new ones. It is also a hope that some of the very hard problems that remain to be solved will ultimately find solutions. Somewhat whimsically, these problems are set apart with the image of a squirrel, indicating that they are "tough nuts to crack".

Certainly the survey is rather personal with a bias towards what has interested me. A large number of new results and examples are also presented. The introduction of the concept of *hyperbolic complete monotonicity* has greatly simplified the exposition and opened new paths.

The work is dedicated to Olof Thorin, whose paper on the Lognormal distribution was my starting point. It took me a month to read that paper but I became fascinated and have been so ever since. Many of the ideas in that paper have been of utmost importance. Thanks go to the departments of mathematical statistics at the Universities of Lund and Stockholm for their hospitality during my sabbatical year of 1987–88 and to the Swedish University of Agricultural Sciences for giving me time to complete the work during spring 1991. I thank C. Berg, K. van Harn, T. Kolsrud, T. Sjödin, F.W. Steutel, and O. Thorin for kind comments and much interest shown in my work. Finally, I thank A. Kaminsky and K. Kaminsky for their mild corrections of my English.

Umeå, January, 1992 Lennart Bondesson

# CONTENTS

# 1. INTRODUCTION AND SUMMARY

## 1.1 Introduction

Various classes of probability distributions can be derived from the Exponential and Gamma distributions by applying different operations, such as convolution, multiplicative convolution, and mixing. The structure and properties of some of these classes are studied in this work. Many very puzzling results appear that still await their strict probabilistic interpretations. Why is a product of independent Gamma variables essentially a sum of independent Gamma variables? Why is the Laplace transform (LT) of a sum of Gamma variables, when considered as a density, the density of a sum of Gamma variables?

Analytic function theory is a tool in this study. Infinite divisibility plays an important role in the exposition but is not intended to be the central theme. However, questions concerning infinite divisibility of various probability distributions have led to many of the results presented here. Since a lot of the standard distributions now are known to be infinitely divisible (ID), the class of ID distributions has perhaps partly lost its interest; smaller classes should be more in focus.

The class of ID distributions was introduced by de Finetti in 1929. In the thirties many beautiful results for this class were derived by different researchers, such as Kolmogorov, Lévy, Khintchine, and others; cf., e.g., Loeve (1978). It was found that the ID distributions precisely constitute the class of limit distributions for triangular null arrays. By the famous Lévy–Khintchine formula, all ID distributions were described in terms of their characteristic function or LT. However, up until 1966 very little was known concerning the ID of many standard distributions given by their densities. In fact, even if the transform is known there can be difficulties in verifying ID. Feller's (1966) paper and the first edition of his (1971) book provided much stimulus for further research. Using different methods, Goldie (1967) and Steutel (1967) obtained the surprising result that mixtures of Exponential distributions are ID; cf. Feller (1971, p. 452). After that there was a period of rather rapid progress. Perhaps the period ended with Steutel's (1973) survey. A new major step forward was taken when Thorin (1977a,b) introduced generalized Gamma convolutions (GGC) and used them to show that in particular the Lognormal distribution is ID. A GGC is the limit distribution for a sequence of sums of independent Gamma variables. A step in the same direction was taken by Grosswald (1976) when proving the ID of the t–distribution. After that many papers appeared and many distributions were identified as being GGC's or at least ID. Perhaps this second period ended in the beginning of the 80's. A survey was published in 1981 (Bondesson, 1981). Steutel (1979) and Lukacs (1983) give very brief reviews.

Reviews in Danish have been given by Berg & Forst (1980) and Ebbesen (1983).

The content of the present work is concentrated on what happened during this second period, but the methods are very much refined and simplified. Moreover, much new material is added.

Some introductory words should be said about the Gamma distribution with density $f(x) = (\Gamma(\beta))^{-1}t^{\beta}x^{\beta-1}e^{-tx}$, $x > 0$, where $\beta$ and t are positive parameters, which is so central here. Although mentioned by Laplace as a natural conjugate prior for the parameter in a Poisson family, this distribution seems to have been in its general form, with $\beta$ real and positive, essentially unknown to the 19th century statistical world. De Forest (cf. Stigler, 1986, pp. 333–4) used it in a demographic study some years before it appeared as a type III distribution in Pearson's (1895) paper. To some extent the late introduction of this distribution is surprising but understandable in view of the strong concentration on error distributions in the 19th century. It could easily have been derived by Gauss if he had also cared to consider scale parameter families for which the sample mean is the best estimator of the parameter, instead of just location parameter families. The properties of the Gamma distribution and those of the Normal distribution resemble each other in many respects. However, it is not so easy to find probability models for which the Gamma distribution naturally appears in its general form. It is symptomatic that neither Johnson & Kotz (1970a), nor Bain (1983), in their surveys mention any such model working for $\beta$ not an integer or half of an integer. Two models are given in this work; they are duals of one another.

## 1.2 Summary

A brief survey of the content is given here. Although the reader is assumed to be familiar with probability theory on a level like that in e.g. Feller's (1971) book, Chapter 2 reviews what is essentially needed. Particular probability distributions and ID distributions are given special attention. Pick functions, i.e., analytic functions with a nonnegative imaginary part in the complex upper half–plane are also treated in this chapter.

Chapters 3–7 constitute the core of the work. In Chapter 3 generalized Gamma convolutions (GGC) are introduced and studied with the help of the moment generating function (mgf) $\varphi(s)$ and Pick function theory. Whereas mixtures of Exponential distributions are characterized by the Pick function property of $\varphi(s)$, the GGC's are characterized by the same property of $\varphi'(s)/\varphi(s)$. A GGC is self–decomposable (SD). Several examples of GGC's are presented. For instance, Stable and $PF_\infty$–distributions on $(0,\infty)$ are GGC's. A composition theorem for GGC's is given; under some conditions on $\zeta(s)$ the function $\varphi(\zeta(s))$ is the mgf of a GGC whenever $\varphi$ is. Two real characterizations of the mgf of a GGC are presented. Another one, far more useful, is given in Chapter 6.

Densities of GGC's and classes of such densities are the focus of Chapter 4. It turns out that a GGC can be represented as a scale mixture of Gamma distributions. Some mixtures of this type are identified as being GGC's. A main result states that a probability density $f(x)$, $x > 0$, which is *hyperbolically completely monotone* (HCM), is the density of a GGC. A function is HCM if, for every $u > 0$, $f(uv)f(u/v)$ is a completely monotone function of $v + v^{-1}$. The class of such densities is denoted $\mathscr{C}$. Thus $\mathscr{C} \subset \mathscr{T}$, where $\mathscr{T}$ is the class of GGC's. This result, perhaps the most important one in this work, has many beautiful and surprising consequences. It has its roots in Thorin's pioneering works. The Lognormal density belongs to $\mathscr{C}$ and so do many other well–known probability densities.

In Chapter 5 the class $\mathscr{C}$ is studied thoroughly. It is shown to coincide with the class of densities which are limits of functions of the form $f(x) = C \cdot x^{\beta-1} \prod_{i=1}^{n}(1+c_i x)^{-\gamma_i}$. The properties of the class are nice. For example, it is closed with respect to multiplication and division of independent random variables and also closed with respect to formation of appropriate powers of random variables. The class is characterized in different ways. The chapter ends with a very hard problem concerning Stable distributions.

In Chapter 6 it is shown that for independent random variables $X \sim \mathscr{T}$, $Y \sim \mathscr{C}$, necessarily $X \cdot Y \sim \mathscr{T}$. This result is a consequence of a real characterization of the LT of a GGC stating that $\phi$ is such a transform iff $\phi(0+) = 1$ and $\phi$ is HCM. A further consequence of that characterization is that the product of a density in $\mathscr{T}$ and the LT of a distribution in $\mathscr{T}$ is essentially a density in $\mathscr{T}$. There are many other consequences. The chapter ends with a discussion on the relation between the HCM–property and logconcavity.

The class $\mathscr{T}_e$ of extended GGC's is considered in Chapter 7. An extended GGC is a limit distribution for linear combinations of independent Gamma variables; the coefficients need not be nonnegative. For instance, all Stable distributions and the t–distribution belong to $\mathscr{T}_e$. Variance mixtures of the Normal distribution with the $\sigma^2$–mixing distribution in $\mathscr{T}$ turns out to be in $\mathscr{T}_e$. Of particular interest is the case when the mixing density is HCM. The corresponding class of distributions on $\mathbb{R}$ is closed with respect to multiplication of independent random variables. The chapter ends with some results concerning self–reciprocal probability densities and with a HCM–formulation of Riemann's hypothesis.

In Chapter 8 the class $\mathscr{T}_d$ of generalized Negative Binomial convolutions (GNBC) is studied. This class is the discrete counterpart of $\mathscr{T}$. It has properties which resemble those of $\mathscr{T}$. Several distributions are shown to belong to $\mathscr{T}_d$. For example, first passage time distributions for simple random walks (with drift) belong to $\mathscr{T}_d$. A large class of GNBC's with explicit probability functions is presented.

Another type of extension of $\mathscr{T}$ is considered in Chapter 9. This new class, $\mathscr{T}_2$, contains all GGC's but also all mixtures of Exponential distributions (MED's) and is closed with respect to convolution and limits. There is a surprisingly simple composition theorem for the $\mathscr{T}_2$–distributions. Several examples of $\mathscr{T}_2$–distributions are presented. For example, first passage time distributions for random walks on $\mathbb{Z}$ in continuous time belong to $\mathscr{T}_2$.

In Chapter 10 shot–noise processes are studied. It turns out that many of the distributions encountered in Chapters 3–9 can be interpreted as marginal distributions for such processes. In particular, the GGC's are obtained when the response function is logconvex and the random effects are exponentially distributed.

The following figures show the relations between some of the classes studied:

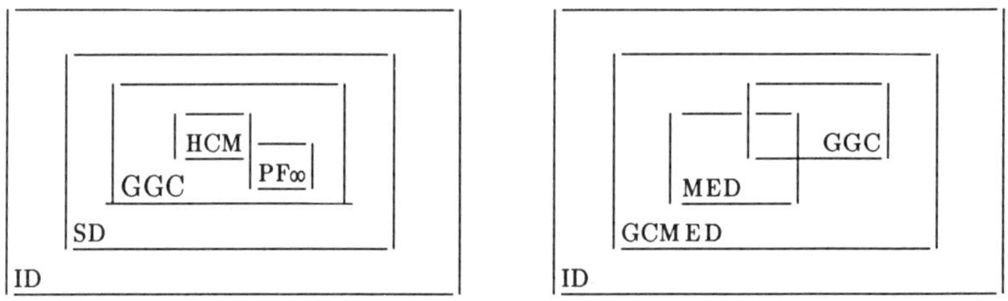


(HCM = $\mathscr{C}$, GGC = $\mathscr{T}$, GCMED = $\mathscr{T}_2$)

## 1.3 General Remarks and Notation

Most chapters end with a collection of miscellaneous results and/or some open problems. Except in Chapter 2, references are in general, but not always, given at the end of the chapter. Most often, references in the main text are inserted to help the reader whereas those at the end of each chapter are more of the credit type. The bibliography is not exhaustive. It contains no textbook on analytic functions; there are many such books.

There are many abbreviations in the text that may frustrate the reader. The abbreviations are explained where they enter and they are later on occasionally reexplained. Both the Laplace transform $\phi(s)$ and the moment generating function $\varphi(s)$ $(= \phi(-s))$ are used in the text. This has been found to be convenient, but the rapid switches between the transforms may at first puzzle the reader.

**List of main symbols and abbreviations:**

| $\mathbb{N}$ | set of natural numbers |
|---|---|
| $\mathbb{N}_0$ | set of nonnegative integers |
| $\mathbb{R}$ | set of real numbers |
| $\mathbb{R}_+$ | set of nonnegative real numbers |
| $\mathbb{C}$ | set of complex numbers |
| $\mathbb{Z}$ | set of integers |
| | |
| $B(\alpha,\beta)$ | Beta function |
| $\Gamma(x)$ | Gamma function |
| $\log z$ | logarithmic function; principal branch |
| $\arg z$ | argument of complex number z |
| C | normalization constant |
| | |
| E | expectation operator |
| Pr | probability |
| $\Pr(\cdot\,\vert\,\cdot)$ | conditional probability |
| X,Y,Z | random variables (rv) |
| $\star$ | convolution |
| f(x) | probability density function (pdf) |
| F(x) | cumulative distribution function (cdf) |
| $\overline{F}(x)$ | survival function |
| $\varphi(s)$ | moment generating function (mgf) |
| $\phi(s)$ | Laplace transform (LT) |
| $p_j$ | probability function (pf) |
| P(z) | probability generating function (pgf) |
| L | Lévy measure |
| U | Thorin measure |
| | |
| Bin(n,p) | Binomial distribution |
| $Po(\lambda)$ | Poisson distribution; mean $\lambda$ |
| $Beta(\alpha,\beta)$ | Beta distribution |
| Exp(t) | Exponential distribution; **mean** $t^{-1}$ |
| $Gamma(\beta,t)$ | Gamma distribution; shape $\beta$, **mean** $\beta/t$ |
| $N(\mu,\sigma^2)$ | Normal distribution |
| $NB(\beta,q)$ | Negative Binomial distribution |
| U(0,1) | Uniform distribution on (0,1) |

| | |
|---|---|
| $\mathscr{P}$ | class of Pick functions |
| $\mathscr{L}$ | class of self–decomposable distributions |
| MED | class of mixtures of Exponential distributions |
| $PF_\infty$ | class of Polya frequency functions |
| $\mathscr{T}$ | class of generalized Gamma convolutions (GGC) |
| $\hat{\mathscr{T}}$ | class of mgf's of GGC's |
| $\check{\mathscr{T}}$ | class of LT's of GGC's |
| $\mathscr{T}_w$ | class of widened GGC's |
| $\mathscr{T}_e$ | class of extended GGC's (EGGC) |
| $\mathscr{T}_d$ | class of generalized Negative Binomial convolutions (GNBC) |
| $\mathscr{C}$ | class of hyperbolically completely monotone densities |
| $\mathscr{B}$ | $\mathscr{B}$–class ($\mathscr{B} = \mathscr{C}$!) |
| iff | if and only if |
| iid | independent and identically distributed |
| a.e. | almost everywhere |
| AM | absolutely monotone |
| CM | completely monotone |
| HCM | hyperbolically completely monotone |
| ID | infinitely divisible (infinite divisibility) |
| SD | self–decomposable |
| SR | self–reciprocal |
| $\sim$ | "distributed as" or "has its distribution in" |
| $\overset{d}{=}$ | equality in distribution |
| $\bar{\ }$ | complex conjugate |
| $\propto$ | proportional to |
| $\sim$ | asymptotic relation |
| $( , )$ | open interval |
| $[ , ]$ | closed interval |
| $\downarrow$ | decreasing (= non–increasing) |
| $\uparrow$ | increasing (= non–decreasing) |

# 2. BASIC CONCEPTS AND FACTS

By and large this chapter, intended for rapid reading, contains only well–known facts from probability theory and analytic function theory. Much of the probability material can be found in e.g. Feller's (1971) book. The chapter serves the purpose of giving the background. It also makes it possible to emphasize certain aspects and to fix the terminology. Pick functions are treated in Section 2.4. Since these functions are not well known to probabilists, the exposition is rather careful but much of the material can also be found in, e.g., Donoghue (1974) or Krein & Nudelman (1977). The chapter ends with some rather recent results concerning infinite divisibility of probability distributions.

## 2.1 Some Facts from Probability Theory

The random variables (rv) $X$, $Y$, $Z$ etc. that are considered are real–valued and most often also nonnegative. The cumulative distribution function (cdf) $\Pr(\{X \leq x\})$ is denoted $F(x)$ or, when necessary, $F_X(x)$. The probability measure is denoted $F(dx)$ and the probability density function (pdf), when it exists, $f(x)$. The transform preferred is the moment generating function (mgf) $\varphi(s) = \mathbf{E}[\exp\{sX\}] = \int e^{sx}F(dx)$. When $X \geq 0$, $\varphi(s)$ is a well–defined analytic function for $\Re s < 0$ taking conjugate values for conjugate points. This transform is used instead of the more conventional Laplace(–Stieltjes) transform (LT) $\phi(s) = \varphi(-s)$ to obtain structural simplicity. However, the LT is used when that is most convenient. The characteristic function equals $\varphi(is)$, $s \in \mathbb{R}$, $i = \sqrt{-1}$, but will be less often used. For a discrete rv $X$, taking values in $\mathbb{N}_0 = \{0,1,2,..\}$, and having probability function $p_j = \Pr(\{X = j\})$, the probability generating function (pgf) $P(z) = \mathbf{E}[z^X] = \Sigma\, p_j z^j$, $0 \leq z \leq 1$, is used.

The transform determines uniquely the probability measure. If that is concentrated on $\mathbb{R}_+ = [0,\infty)$, even the values of the mgf in an interval $(-\epsilon,0)$ determine it. Addition of independent rv's corresponds to convolution, $\star$, of the associated probability measures (pdf's) and to multiplication of the transforms.

A sequence $\{F_n(dx)\}_{n=1}^{\infty}$ of probability measures is said to converge weakly to another probability measure $F(dx)$ if

$$(2.1.1) \qquad \int g(x)F_n(dx) \to \int g(x)F(dx) \quad \text{as } n \to \infty,$$

for all continuous functions $g$ with finite limits as $x \to \pm\infty$. A corresponding sequence of rv's is said to converge in distribution. Weak convergence implies that $F_n(x) \to F(x)$ at all continuity points x of F and also that (2.1.1) holds for all bounded continuous functions. Thus, when there is weak convergence, the sequence of transforms converges pointwise.

When the probability measures $F_n(dx)$ have their masses in a compact part of $\mathbb{R}$ or compact extension of $\mathbb{R}$, the sequence can also be thinned to converge weakly to a probability measure; this is essentially Helly's first theorem. Letting for simplicity the original space be $\mathbb{R}_+$ and extending it to $[0,\infty]$, we have a compact space. Hence the sequence $F_n(dx)$ can be thinned to converge weakly to a probability measure $F(dx)$ that, however, might have positive mass at $\infty$. The corresponding mgf is given by

$$\varphi(s) = \int_{[0,\infty]} e^{sx}F(dx) = \begin{cases} 1\,, & s = 0 \\ \int_{[0,\infty)} e^{sx}F(dx), & s < 0 \end{cases}$$

If the sequence of mgf's $\varphi_n(s)$ converges pointwise to some function $\varphi(s)$, then by the uniqueness of the transform, there is only one possible weak limit, and it follows that the sequence $F_n(dx)$ converges weakly to a probability measure $F(dx)$ with mgf $\varphi(s)$. The probability measure $F(dx)$ is non–defective, i.e. has no mass at $\infty$, iff $\lim_{s\to 0} \varphi(s) = 1$.

For the space $\mathbb{R}$, the characteristic function $\varphi(is)$, $s \in \mathbb{R}$, is used instead. The conclusions above are not changed but somewhat more complicated to derive since $e^{isx}$ has no limit as $x \to \pm\infty$.

Unless otherwise stated, the probability measures we consider have no mass at $\infty$. A well–known theorem of Scheffé asserts that if $\{f_n\}$ is a sequence of pdf's that converges pointwise (a.e.) to a pdf $f(x)$, then the corresponding sequence of probability measures $F_n(dx)$ converges weakly towards $F(dx)$. Occasionally we consider also nonnegative measures $\mu(dx)$ on e.g. $\mathbb{R}$ or $\mathbb{R}_+$ that are not probability measures. A sequence $\{\mu_n\}$ of such measures is said to converge vaguely to $\mu$ if $\int g(x)\mu_n(dx) \to \int g(x)\mu(dx)$ for all continuous functions $g(x)$ with compact support in the space considered and thus for all continuous functions if the space itself is compact. If the sequence of total masses $\int \mu_n(dx)$ is bounded, the sequence $\{\mu_n\}$ can be thinned to converge vaguely.

A well–known characterization of the LT of a nonnegative measure on $\mathbb{R}_+$ is *Bernstein's theorem.* It states that a function $\phi$ on $(0,\infty)$ (or $(a,\infty)$ ) is of the form $\phi(s) = \int_{[0,\infty)} e^{-sx}\mu(dx)$ with $\mu$ nonnegative iff $\phi$ is completely monotone (CM), i.e. $(-1)^n\phi^{(n)}(s) \geq 0$, $n \in \mathbb{N}_0$. (Certainly $\phi(s) \to 1$ as $s \to 0$ iff $\mu$ is a probability measure.) Complete monotonicity is preserved under mixing (i.e., formation of nonnegative linear combinations), multiplication and pointwise limits. If $\xi(s)$ is nonnegative with a CM derivative, then $\phi(\xi(s))$ is CM whenever $\phi$ is. For instance, $\exp\{-s^\alpha\}$ and $1/(1+s^\alpha)$, $0 \le \alpha \le 1$, are CM as a consequence of that result.

A discrete counterpart of Bernstein's theorem states that a function $P(z)$ on $[0,1)$ is absolutely monotone (AM), i.e., $P^{(n)}(z) \ge 0$, $n \in \mathbb{N}_0$, iff it can be represented as $P(z) = \Sigma_{j=0}^{\infty} p_j z^j$, where the $p_j$'s are nonnegative. Thus if $P$ is AM and $P(z) \to 1$ as $z \to 1$, then $P$ is a probability generating function.

Another discrete version of Bernstein's theorem is *Hausdorff's characterization* of CM sequences, i.e. sequences $\{r_j\}_0^\infty$ for which $(-1)^n\Delta^{(n)}r_j \ge 0$, $j \in \mathbb{N}_0$, $n \in \mathbb{N}_0$. Here $\Delta^{(n)}r_j = \Delta^{(n-1)}r_{j+1} - \Delta^{(n-1)}r_j$ and $\Delta^{(0)}r_j = r_j$. The characterization states that $\{r_j\}_0^\infty$ is CM iff it admits the representation $r_j = \int_{[0,1]} p^j H(dp)$, where $H$ is a nonnegative measure.

A less restrictive condition than complete monotonicity is *logconvexity*. A positive function $f$, possibly a pdf, is logconvex if $\log f$ is convex. Logconvexity is preserved under pointwise limits, multiplication, and the operation $f \to f^\gamma$, $\gamma \ge 0$. More surprisingly, it is preserved under mixing as well. This is *Artin's theorem*; see, e.g., Marshall & Olkin (1979, p. 452). If $f$ is logconvex, then so is $\overline{F}(x) = \int_x^\infty f(y)dy = \int_0^\infty f(x+y)dy$ as is immediate from that theorem.

While logconvexity is preserved under mixing, *logconcavity* is preserved under convolution; see, e.g., Karlin (1968, p. 333 and Chapt. 3), or Dharmadhikari & Joag–dev (1988, pp. 17–23). If $f$ is logconcave, so is $\overline{F}$. In fact, $\overline{F}$ is the convolution of $f$ and the logconcave indicator function of the interval $(-\infty, 0)$. Logconcave densities are often called *strongly unimodal* since $f \star g$ is unimodal whenever $g$ is if and only if $f$ is logconcave (Ibragimov, 1956). The Gamma density is logconvex and CM for $\beta \le 1$ and logconcave for $\beta \ge 1$.

## 2.2 Particular Probability Distributions

Some familiar probability distributions are discussed in this section. An attempt is made to concentrate more on less well–known facts than on facts known from standard expositions, such as, e.g., Johnson & Kotz (1969, 1970a, 1970b).

***Poisson distribution and Poisson process.*** A discrete rv $X$ is Poisson distributed with mean $\lambda$ if

$$p_j = \Pr(\{X=j\}) = \frac{\lambda^j}{j!}e^{-\lambda}, \quad j \in \mathbb{N}_0.$$

The distribution is denoted $Po(\lambda)$. Its pgf is $P(z) = \exp\{\lambda(z-1)\}$. The distribution appears as the limit distribution of the Binomial distribution $Bin(n,p)$ as $n \to \infty$, $p \to 0$, and $np \to \lambda$ and has found many applications. If the real line is split into small intervals of length $h$ and in each a point is put with probability $p = \lambda \cdot h$, we get as $h \to 0$ a point process called the Poisson process of intensity $\lambda$. The number of points in an interval (or set) $I$ of length $|I|$ is $\mathrm{Po}(\lambda|I|)$–distributed. The numbers of points in disjoint sets are independent.

***Exponential and Geometric Distributions.*** An rv $X \geq 0$ is exponentially distributed if it has a pdf of the form $f(x) = te^{-tx}$, $x \geq 0$, where $t > 0$. The distribution is denoted $\mathrm{Exp}(t)$. Its mgf is $\varphi(s) = \frac{t}{t-s} = \frac{1}{1 - s/t}$. As $t \to \infty$, the distribution tends to a one–point distribution at $0$. For a Poisson process of intensity $\lambda = t$, the intervals between successive points have lengths distributed as $\mathrm{Exp}(t)$. A characteristic property of the Exponential distribution is its lack of memory, i.e., $\Pr(\{X-x \leq y \mid X > x\}) = \Pr(\{X \leq y\})$. Clausius (1858) might have been the first one to use the Exponential distribution in connection with free path lengths for gas molecules.

The discrete analogue of the Exponential distribution is the Geometric distribution with probability function $p_j = pq^j$, $j \in \mathbb{N}_0$ $(0 \leq q < 1,\ p = 1-q)$, and pgf $P(z) = \frac{p}{1-qz} = (1 - \frac{q}{p}(z-1))^{-1}$. The number of failures in a binomial experiment until there is success the first time is geometrically distributed. The pgf equals $\varphi(z-1)$, where $\varphi$ is the mgf of $\mathrm{Exp}(t)$ with $t = p/q$. The interpretation is that the Geometric distribution is a mixture of $\mathrm{Po}(\lambda)$–distributions with $\lambda \sim \mathrm{Exp}(t)$. The Exponential and Geometric distributions appear often rather unexpectedly. The following little known example illustrates that.

**Example 2.2.1** (***Clumping of particles***). Consider a very large population of particles which at time $\tau = 0$ all have mass (size) 1. Then they start colliding completely randomly with each other and merge pairwise into new particles. In the time interval $(\tau, \tau+\Delta\tau)$ a fraction $2\Delta\tau$ of the particles are involved in the process. Let $P(z;\tau)$ denote the pgf of the distribution of the size of the particles at time $\tau$. Then

$$P(z;\ \tau+\Delta\tau) = \frac{(1-2\Delta\tau)P(z;\tau) + (P(z;\tau))^2\Delta\tau}{1-\Delta\tau}.$$

The denominator adjusts for the reduced number of particles. Hence $P^{-1}(1-P)^{-1}\partial P/\partial\tau = -1$. By a partial fraction decomposition and then an integration,

$$P(z;\ \tau) = \frac{P(z;0)e^{-\tau}}{1 - (1-e^{-\tau})P(z;0)}. \tag{2.2.1}$$

Thus, for $P(z;0) = z$, a Geometric distribution shifted one step to the right appears.

When the initial size distribution is not discrete and has LT $\phi(s; 0)$, (2.2.1) holds analogously for $\phi(s; \tau)$. Assuming that $\phi(s; 0) \sim 1 - as^{\alpha}$, $0 \le \alpha \le 1$, $a \ge 0$, as $s \to 0$ and rescaling the size by a factor $c(\tau) = e^{-\tau/\alpha}$, i.e., replacing $\phi(s; \tau)$ by $\phi(c(\tau)s; \tau)$, we get $\phi(c(\tau)s; \tau) \sim (1+as^{\alpha})^{-1}$ asymptotically as $\tau \to \infty$. In particular the Exponential distribution appears as a limit distribution if the initial distribution has finite mean. The model here has been used to model the size of raindrops and in connection with cell fusion; see, e.g., Scott (1968) and Röhme & Thorburn (1981). □

***Gamma distribution.*** The sum of n independent Exp(t)–variables has the pdf and mgf

$$f(x) = \frac{t^n}{\Gamma(n)} x^{n-1} e^{-tx}, \quad x > 0, \qquad \varphi(s) = \left(\frac{t}{t-s}\right)^n,$$

respectively. Replacing n by any real number $\beta > 0$, we get the general Gamma($\beta$, t) distribution. For t fixed the family is closed under convolution:

$$\text{Gamma}(\beta_1, t) \star \text{Gamma}(\beta_2, t) = \text{Gamma}(\beta_1+\beta_2, t)\ .$$

There is a similarity between the pdf and the mgf. It is most apparent when a location parameter is also introduced. The pdf $f(x) = C\cdot(x-a)^{\beta-1}e^{-tx}$, $x > a$, has the mgf $C'\cdot(t-s)^{-\beta}e^{as}$. This similarity may partially explain the many curious results that appear later on.

Although the general Gamma distribution has been much used in applied model–building, it is rarely encountered in more theoretical probability models. One example of its appearance is presented below.

**Example 2.2.2** (***Random reduction***). Let $V_1, V_2, \ldots$ be rv's with the Exp(1)–distribution and $U_1, U_2, \ldots$ rv's with the uniform distribution U(0,1). All the rv's are independent. A sequence $X_1, X_2, \ldots$ of new rv's is defined recursively by

$$X_n = U_n^{\gamma}\cdot(V_n + X_{n-1}), \quad n = 1, 2, \ldots, \quad \gamma > 0, \tag{2.2.2}$$

where $X_0$ is defined arbitrarily. We may think of $V_n$ as the time between two successive epochs $T_{n-1}$ and $T_n$ in a Poisson process and $X_{n-1}$ as the size, just after $T_{n-1}$, of a population that increases linearly in time. At time $T_n$ the size is reduced by a factor $1 - U_n^{\gamma}$; the pdf of $U_n^{\gamma}$ is $\beta\cdot x^{\beta-1}$, $0 < x < 1$, with $\beta = 1/\gamma$. Asymptotically the rv's $X_n$ have a common distribution. Let $\phi(s)$ be the LT of that distribution. Then, from (2.2.2),

$$\phi(s) = \int_0^1 \frac{1}{1+sx}\, \phi(sx)\, \beta x^{\beta-1} dx\ .$$

Substituting $sx = y$, multiplying then both sides by $s^\beta$ and differentiating $g(s) = s^\beta\phi(s)$ with respect to $s$, we get

$$g'(s)/g(s) = \beta\cdot\frac{1}{1+s}\cdot\frac{1}{s} = \beta\cdot(\frac{1}{s} - \frac{1}{1+s}),$$

and hence, by an integration, $g(s) = C\cdot(\frac{s}{1+s})^\beta$, where C is a constant. Since $\phi(0) = 1$, it follows that $\phi(s) = (1+s)^{-\beta}$, i.e., the asymptotic distribution is Gamma($\beta$, 1).

The model provides a model–based simple method of generating approximative random numbers from the general Gamma distribution. The methods used in practice are artificial. The model is old; Feller (1971, pp. 325–326) mentions it as a model for stellar radiation. Grandell (1985, pp. 72–73) uses it as a model for air pollutant concentration. The model is encountered again in Chapter 10 in a dual form. □

***Negative Binomial distribution.*** This distribution, here denoted NB($\beta$, q), has probability function and pgf as follows

$$p_j = (-1)^j\binom{-\beta}{j} q^j p^\beta, \quad j \in \mathbb{N}_0, \qquad P(z) = (\frac{p}{1-qz})^\beta \qquad (0 \le q < 1;\ p = 1-q;\ \beta > 0).$$

The pgf equals $\varphi(z-1)$, where $\varphi$ is the mgf of a Gamma($\beta$, t)–distribution with $t = p/q$. Thus the NB–distribution is a Po($\lambda$)–mixture with $\lambda \sim$ Gamma($\beta$, t); Greenwood & Yule (1920). The NB–distribution appears in many probability models; see, e.g., Boswell & Patil (1970) or Tripathi (1985). A less well–known example of its appearance is now given. It is similar to Example 2.2.2.

**Example 2.2.3** (***Random thinning***). Set

$$X_n = \sum_{i=1}^{X_{n-1}+V_n} Z_{in}, \qquad n = 1, 2, \ldots,$$

where all the rv's on the right hand side are independent. We let $V_n \sim NB(1,q)$ (Geometric distribution) and $Z_{in} \sim Bin(1,\pi)$ with $\pi$ random and $\pi = U_n^\gamma$, $U_n \in U(0, 1)$, $\gamma > 0$. The asymptotic distribution of $X_n$ has a pgf satisfying the relation

$$P(z) = \int_0^1 \psi(1-\pi+\pi z)\ \beta\cdot\pi^{\beta-1}\, d\pi,$$

where $\psi(z) = P(z)\cdot\frac{p}{1-qz}$ and $\beta = 1/\gamma$. Substituting $y = \pi(1-z)$, multiplying both sides by $(1-z)^\beta$, and then differentiating $g(z) = (1-z)^\beta P(z)$ with respect to $z$, we get an equation that easily gives $P(z) = (\frac{p}{1-qz})^\beta$, i.e., the NB($\beta$, q)–distribution.

This example is a discrete variant of Example 2.2.2. The continuous random reduction there is replaced by a random thinning. A dual form of this example is encountered in Chapter 10. □

***Related distributions.*** Several distributions can be derived from the Gamma distribution by simple operations.

Letting $X \sim \text{Gamma}(\beta, t)$ and considering $Y = X^{1/\alpha}$, $\alpha \neq 0$, we have

$$f_Y(x) = \frac{|\alpha|}{\Gamma(\beta)} t^\beta x^{\beta\alpha-1} \exp\{-tx^\alpha\}, \quad x > 0; \tag{2.2.3}$$

a *generalized Gamma distribution*. For $\alpha = -1$ $(Y = X^{-1})$, the distribution is sometimes called an *inverse Gamma distribution*. Multiplying this particular pdf by a factor $\exp\{-c_1x\}$, and changing $\beta$ to $-\beta$, a density of the form

$$f(x) = C \cdot x^{\beta-1} \exp\{-c_1x - c_2 \tfrac{1}{x}\}, \quad x > 0 \quad (c_1, c_2 > 0) \tag{2.2.4}$$

appears. We may here permit $\beta$ to be any real number. These distributions are called *generalized inverse Gaussian distributions* and have been studied by Jorgensen (1981). For $\beta = -1/2$ they reduce to the inverse Gaussian distributions.

Looking again at (2.2.3) and expanding $x^\alpha = \exp\{\alpha \cdot \log x\}$ in a McLaurin series with respect to $\alpha$, we find, after some calculation, that as $\alpha \to 0$, $\beta \to \infty$, $t \to \infty$, $\alpha^2 t \to 1/\sigma^2$, $t\alpha - \beta\alpha \to \mu/\sigma^2$, the *Lognormal density*

$$f(x) = \frac{1}{\sqrt{2\pi}} \frac{1}{\sigma} \frac{1}{x} \exp\{-\frac{(\log x - \mu)^2}{2\sigma^2}\}.$$

appears as a limit distribution; cf. Ferguson (1962). It corresponds to $Y = e^X$, where $X \sim N(\mu, \sigma^2)$.

The ratio $X = X_1/X_2$ of two independent rv's $X_j \sim \text{Gamma}(\beta_j, 1)$, $j = 1, 2$, has the pdf

$$f_X(x) = \int_0^1 y \cdot f_{X_1}(xy) f_{X_2}(y) dy = \frac{1}{B(\beta_2, \beta_1)} \cdot x^{\beta_1 - 1}/(1+x)^{\beta_1+\beta_2}; \tag{2.2.5}$$

a *Beta-distribution of the second kind* with unit scale. The pdf of the rv $Y = X_1/(X_1+X_2) = X/(1+X)$ equals

$$f_Y(x) = \frac{1}{B(\beta_2, \beta_1)} \cdot x^{\beta_1-1}(1-x)^{\beta_2-1}, \quad 0 < x < 1.$$

This is the *Beta-distribution*. It is well known that $Y$ is independent of

$X_1+X_2 \sim \text{Gamma}(\beta_1+\beta_2, 1)$. Since the distribution of $(Y, X_1+X_2)$ determines that of $(X_1, X_2)$, we have, somewhat informally expressed,

$$\text{Gamma}(\beta_1, 1) = \text{Beta}(\beta_1, \beta_2) \cdot \text{Gamma}(\beta_1+\beta_2, 1) ,$$

as found by Stuart (1962). In fact, this relation was encountered in Example 2.2.2 for $\beta_1 = \beta$ and $\beta_2 = 1$.

The density (2.2.5) can be specialized and extended. Letting $\beta_1 = 1$, we have the (modified) *Pareto distribution* (with unit scale) with pdf

$$f(x) = \beta_2 \cdot (1+x)^{-\beta_2-1} = \int_0^\infty te^{-tx} \cdot \frac{1}{\Gamma(\beta_2)} \cdot t^{\beta_2-1} e^{-t} dt.$$

The integral shows that $f$ can be interpreted as the pdf of an Exp(t) variable for which $t$ is random and $t \sim \text{Gamma}(\beta_2, 1)$. An extension of (2.2.5) is provided by

$$f(x) = \frac{|\alpha|}{B(\beta_2, \beta_1)} \cdot x^{\beta_1\alpha-1} \cdot (1+x^\alpha)^{-\beta_1-\beta_2}$$

that corresponds to the power $1/\alpha$ of the ratio $X_1/X_2$ of two Gamma variables.

One may also consider a product $X = X_1 \cdot X_2$ of two independent rv's $X_j \sim \text{Gamma}(\beta_j, 1)$, $j = 1, 2$. Its pdf is

$$f_X(x) = \int_0^\infty \frac{1}{y} \cdot f_{X_1}(x/y) \cdot f_{X_2}(y) dy = \frac{1}{\Gamma(\beta_1)\Gamma(\beta_2)} \cdot x^{\beta_1-1} \int_0^\infty y^{\beta_2-\beta_1-1} \exp\{-\frac{x}{y} - y\}\, dy .$$

The expression is inexplicit. The substitution $y' = y^{-1}$ makes it only slightly simpler.

***Distributions with given Laplace transforms.*** We now turn to some distributions that are defined in terms of their Laplace transforms and in a certain sense are dual to some of the distributions encountered above.

A nonnegative rv has a (strictly) *Stable distribution* with index $\alpha$, $0 < \alpha < 1$, if its LT is of the form $\phi(s) = \exp\{-cs^\alpha\}$. Here c, or rather $c^{1/\alpha}$, is a scale parameter. A sum of independent Stable rv's of index $\alpha$ is a Stable rv of index $\alpha$. There is also a multiplicative property: If $X_\alpha$ denotes a Stable rv of index $\alpha$ and unit scale, then, under independence, $X_\beta^{1/\alpha} \cdot X_\alpha$ has a Stable distribution of index $\alpha\beta$. This is verified by a use of conditional expectations:

$$E[\exp\{-sX_\beta^{1/\alpha}X_\alpha\}] = E[E[\exp\{-sX_\beta^{1/\alpha}X_\alpha\} \,|\, X_\beta]] = E[\exp\{-s^\alpha X_\beta\}] = \exp\{-s^{\alpha\beta}\}.$$

Analogously, the LT of the rv $Y^{1/\alpha}\cdot X_\alpha$, where $Y \sim \text{Gamma}(\gamma, 1)$, equals $\phi(s) = (1+s^\alpha)^{-\gamma}$. For $\gamma = 1$ this LT was encountered in Example 2.2.1.

There is only an explicit expression for the pdf of a Stable distribution when $\alpha = 1/2$. For $\alpha = 1/2$ and $c = 1$, the pdf is

$$f(x) = \frac{1}{2\sqrt{\pi}}\cdot x^{-3/2}\exp\{-\tfrac{1}{4x}\}, \quad x > 0,$$

and happens to be a special inverse Gamma distribution. It appears as a passage time distribution for Brownian motion without drift.

## 2.3 Infinite Divisibility

A probability distribution F is said to be infinitely divisible (ID) if, for each $n \geq 1$, it can be decomposed into n identical convolution factors $F_n$. The mgf $\varphi(s)$ can then be written $\varphi(s) = (\varphi_n(s))^n$, where $\varphi_n(s)$ is the mgf of a probability distribution. We talk also about ID rv's and mgf's. The Gamma, Negative Binomial, and Stable distributions are ID. The mgf of an ID distribution on $\mathbb{R}_+ = [0,\infty)$ can be represented canonically as

$$\text{(2.3.1)} \qquad \varphi(s) = \exp\{as + \int_{(0,\infty)}(e^{sy}-1)L(dy)\}\ ,$$

where $a \geq 0$ and L, *the Lévy measure,* is nonnegative and satisfies $\int_{(0,\infty)}\min(1,y)\, L(dy) < \infty$. This latter condition is necessary and sufficient for the convergence of the integral in the exponent. The parameter $a$ is the left–extremity of the distribution. Also L is uniquely determined. The representation above is essentially a special case of the more general *Lévy(–Khintchine) representation* of the mgf of any ID distribution on $\mathbb{R}$:

$$\text{(2.3.2)} \qquad \varphi(s) = \exp\{as + \frac{\sigma^2}{2}\cdot s^2 + \int_{y\neq 0}(e^{sy} - 1 - \frac{sy}{1+y^2})L(dy)\}\ , \quad \Re e\, s = 0.$$

where $a \in \mathbb{R}$ and $\int_{y\neq 0}\min(1,y^2)\, L(dy) < \infty$. This representation is not used much here.

The class of ID distributions is closed under convolution and weak limits. The ID distributions also appear as the class of possible limit distributions for sums $S_n = \Sigma_{j=1}^n X_{nj}$, where the rv's are independent but not necessarily identically distributed and satisfy some asymptotic negligibility condition of the type $\lim_{n\to\infty} \sup_j \Pr(\{|X_{nj}| > \epsilon\}) = 0$. The ID distributions also appear in other more or less related contexts. Their appearance in connection with passage times for skip–free

Markov processes and shot–noise processes is treated in Section 9.3 and Chapter 10. They also appear in potential theory, see, e.g., Berg & Forst (1979), and in renewal theory, see, e.g., Kingman (1972), van Harn (1978), and Hansen (1988). More abstract infinite divisibility is treated in Kendall & Harding (1973) and Ruzsa & Szekely (1988).

A *Lévy process* is a random process $\{X(\tau),\ \tau \geq 0\}$, with stationary independent increments and such that, when the increments are nonnegative,

$$\varphi_{X(\tau)}(s) = \exp\{\tau\cdot(as + \int_{(0,\infty)}(e^{sy}-1)L(dy))\}.$$

The Lévy measure has a nice interpretation. For simplicity, let $a = 0$. Assume first that $\lambda = \int_{(0,\infty)}L(dy) < \infty$. Let $N(\tau)$ be the number of points in the interval $(0,\tau]$ of a Poisson point process of intensity $\lambda$ and let $Y_j$, $j = 1, 2,\dots$, be iid rv's with probability measure $L(dy)/\lambda$. The *compound Poisson process* $X(\tau) = \Sigma_{j=1}^{N(\tau)} Y_j$ is a Lévy process with Lévy measure L. The $Y_j$'s are lenghts of jumps taking place at the epochs of the Poisson point process. When $\lambda = \infty$, we may consider a sequence of measures $L_n(dy)$, $n = 0, 1, 2,\dots$, such that $L_0(dy)$ equals $L(dy)$ on $(1,\infty)$ and is zero elsewhere, whereas, for $n \geq 1$, $L_n(dy)$ equals $L(dy)$ on $(2^{-n},2^{-n+1}]$ and is zero elsewhere. Forming the sum of the corresponding independent compound Poisson processes, we get a Lévy process with given Lévy measure. The process has infinitely many jumps in each interval but the sum of the lengths of the jumps is finite. The epochs of jumps with length exceeding x form a Poisson point process of intensity $\lambda_x = \int_{(x,\infty)}L(dy)$. This representation is known as the *Lévy–Ito representation.* One may think of the Lévy process as a cumulative damage process.

***Verification of ID.*** Though (2.3.1) gives a complete description of the mgf of all ID distributions on $\mathbb{R}_+$, it is often hard to verify that a given distribution on $\mathbb{R}_+$ is ID. There can be difficulties even if the mgf is known.

Certainly (2.3.1) is equivalent to

$$\text{(2.3.3)} \qquad \frac{d}{ds}\log \varphi(s) = \frac{\varphi'(s)}{\varphi(s)} = a + \int_{(0,\infty)} e^{sy}yL(dy) = \int_{[0,\infty)}e^{sy}R(dy),$$

where the measure R equals $yL(dy)$ for $y > 0$ and has mass $a$ at $y = 0$. Thus ID is equivalent to absolute monotonicity (AM) of $\varphi'(s)/\varphi(s)$ by Bernstein's theorem. For the Gamma($\beta$, t) distribution we have $\varphi'(s)/\varphi(s) = \frac{\beta}{t-s}$ and the R–measure has the exponential density $\beta e^{-ty}$, $y > 0$. For other distributions, however, it might be more difficult to verify AM. When $a = 0$ and the given distribution has a pdf $f(x)$ on $(0,\infty)$, we may reformulate (2.3.3) as

$$(2.3.4) \qquad xf(x) = \int_{(0,x]} f(x-y)R(dy)\,, \quad x > 0.$$

As observed by Steutel (1973), ID holds iff there is a nonnegative solution $R(dy)$ to this integral equation.

***Creation of new ID distributions.*** There is a simple way of composing two ID mgf's to a new one. Let $\exp\{\zeta(s)\}$ be the mgf of an ID distribution (maybe on $\mathbb{R}$) and let $\varphi(s)$ be the mgf of an ID distribution on $\mathbb{R}_+$. Then $\varphi(\zeta(s))$ is also ID. This can be seen as follows. Let $X(\tau)$, $\tau \geq 0$, be a Lévy process with mgf $\exp\{\tau\zeta(s)\}$ and $T \geq 0$ an independent random time with mgf $\varphi(s)$. The mgf of $X(T)$ is $\varphi(\zeta(s))$. We have to show that $X(T)$ is ID. We only check that $X(T)$ can be written as a sum of two iid rv's. Let $T = T_1+T_2$, where $T_1$ and $T_2$ are iid. Now

$$X(T) = [X(T_1+T_2) - X(T_1)] + X(T_1)\,.$$

For fixed values of $T_1$ and $T_2$, the two terms on the right hand side are independent and the first term is distributed as $X(T_2)$. Some reflection then shows that this conditional independence implies independence. Thus the two terms are iid.

Since $\varphi(\zeta(s)) = \int_{[0,\infty)} \exp\{\tau\zeta(s)\}F_T(d\tau)$, a *convolution mixture* of an ID distribution with an ID mixing distribution is ID. There are many applications of this result; see, e.g., Kent (1981).

An extremely simple way of forming a new ID distribution from an old one is to multiply the probability measure $F(dx)$ by an exponential factor $e^{-\theta x}$ and then normalize to get a probability measure. Mathematically the ID is more or less a triviality. The operation is simplest to interpret probabilistically when $\theta \geq 0$ and $F(dx)$ is concentrated on $\mathbb{R}_+$, and that interpretation is given here. Let $X \sim F$ and let $Y \sim \mathrm{Exp}(\theta)$ be independent of $X$. Set

$$\tilde{X} = \begin{cases} X & \text{if } X \leq Y \\ \infty & \text{if } X > Y \end{cases}$$

We may think of $X$ as a life–length and $Y$ as an emigration time. The variable $X$ is observed only if death occurs before emigration. Otherwise it is recorded as $\infty$. Now if $X$ is a sum of iid rv's $X_1, \ldots, X_n$, then $\tilde{X}$ is the sum of the iid rv's $\tilde{X}_1, \ldots, \tilde{X}_n$ defined as $\tilde{X}$ with different independent $Y$'s $\sim \mathrm{Exp}(\theta)$. It is the lack of memory of the Exponential distribution that makes the variables $\tilde{X}_j$ independent. In an obvious sense the defective probability measure $e^{-\theta x}F(dx)$ is now ID and hence so is the corresponding probabability measure. Later it will be seen that under certain conditions the result sometimes also holds when $\theta$ is an rv. That is difficult to explain probabilistically.

***Self–decomposability.*** A more restrictive concept than ID is self–decomposability (SD). An rv is said to be self–decomposable (SD) if, for each c, $0 < c \leq 1$, we have

$$X \stackrel{d}{=} c \cdot X + \epsilon_c,$$

where $\epsilon_c$ is an rv independent of X and $\stackrel{d}{=}$ means equality in distribution. Equivalently, a probability distribution with mgf $\varphi(s)$ is SD if, for each c, $0 < c \leq 1$, $\varphi(s)/\varphi(cs)$ is the mgf of a probability distribution. The SD distributions were introduced by Lévy as the limit distributions for normed partial sums of a sequence of independent rv's; see, e.g., Feller (1971, pp. 588–589) or Petrov (1975, pp. 82–87). The class of SD distributions, also called the $\mathscr{L}$–class, is a subclass of the ID distributions and is closed under convolution and weak limits. A non–degenerate SD distribution is absolutely continuous and unimodal; see Yamazato (1978), Lukacs (1983, p. 48 and pp. 57–64), or Dharmadhikari & Joag–dev (1988, pp. 122–140).

If $X_0$ is SD, then we may define for arbitrary c, $0 < c \leq 1$, a stationary autoregressive process of order 1 by setting

$$X_{n+1} = cX_n + \epsilon_n, \quad n \in \mathbb{N}_0,$$

and by choosing the distribution of the iid rv's $\epsilon_n$ appropriately. If $N(\tau)$ is a Poisson process or more generally a Lévy process, then, for fixed $\rho > 0$, the process

$$X(\tau) = \int_{(-\infty,\tau]} \exp\{-\rho\cdot(\tau-\tau')\} \, N(d\tau')$$

is SD in the sense that, for each $h > 0$, $X(\tau+h) = e^{-\rho h} \cdot X(\tau) + \epsilon$, where $X(\tau)$ and $\epsilon$ are independent; cf., e.g., Vervaat (1979).

The SD distributions are characterized as follows:

*An ID distribution on $\mathbb{R}_+$ with mgf $\varphi$ is SD iff its Lévy measure L has a density $\ell$ such that $r(y) = y\ell(y)$ is decreasing.*

(In the literature this result is most often formulated in a more complicated way; cf., e.g. Lukacs (1983, p. 47).) In fact, if r(y) is decreasing then, as is easily verified, the mgf $\varphi(s)/\varphi(cs)$, corresponding to $\epsilon_c$, is ID with a Lévy measure that has the density $\ell(y) - c^{-1}\ell(y/c)$. The necessity of the condition for SD is more difficult to verify as the following sketch shows. Consider the tail $\overline{L}(x) = L((x,\infty))$. The corresponding tail of the Lévy measure $L_c$ of $\epsilon_c$, which can be verified to be ID, equals $\overline{L}_c(x) = \overline{L}(x) - \overline{L}(x/c)$ and is as a tail decreasing. Setting $\overline{M}(x') = \overline{L}(e^{x'})$, we find that $\overline{M}(x') - \overline{M}(x'+h)$ is decreasing for any fixed $h > 0$ ($c^{-1} = e^h$). It follows that $\overline{M}(x)$ is convex and thus it can be represented as an integral from x' to $\infty$ of a decreasing function m(y'), i.e., $\overline{M}(x') = \int_x^\infty m(y')dy'$. Transforming back to $\overline{L}$, we have $y \cdot \ell(y) = m(\log y)$ which is decreasing as desired.

The Gamma($\beta$, t) distribution is SD since $r(y) = y\ell(y) = \beta\cdot e^{-ty}$ is decreasing; r is even completely monotone. For the Stable distribution (with unit scale) we have $\varphi(s) = \exp\{-(-s)^\alpha\}$ and thus it is SD since

$$\varphi'(s)/\varphi(s) = \alpha(-s)^{\alpha-1} = \int_0^\infty e^{sy} r(y)dy ,$$

and $r(y) = \frac{\alpha}{\Gamma(1-\alpha)} y^{-\alpha}$ is decreasing. The Stable distribution has, contrary to the Gamma distribution, the property of being in Urbanik's class $\mathscr{L}^\infty$ of *completely SD distributions:*

Let $\mathscr{L} = \mathscr{L}^0$ be the class of SD distributions. Defining then iteratively $\mathscr{L}^k$ $(k = 1, 2, \ldots)$ as the class of distributions (of X) in $\mathscr{L}^{k-1}$ for which $\epsilon_c \sim \mathscr{L}^{k-1}$, we have $\mathscr{L}^\infty = \cap_{k=0}^\infty \mathscr{L}^k$. Urbanik (1972) has shown that $\mathscr{L}^\infty$ equals the class of limit distributions of triangular arrays of independent Stable rv's (of different indices $\alpha$). The mgf is of the form

$$\varphi(s) = \exp\{-\int_{(0,1]}(-s)^\alpha K(d\alpha)\},$$

where K is a nonnegative measure. See also Kumar & Schreiber (1978) and Berg & Forst (1983).

If $X(\tau)$ is a Stable Lévy process (of index $\alpha$) and T is an independent SD rv, then X(T) is SD. In fact

$$X(T) \stackrel{d}{=} X(cT+\epsilon_c) = X(cT) + [X(cT+\epsilon_c) - X(cT)].$$

Here the two main terms are independent, which we have noticed before. The first term is distributed as $c^{1/\alpha}X(T)$ and the second term as $X(\epsilon_c)$, which gives the desired result. Equivalently, $\varphi(-(-s)^\alpha)$ is the mgf of an SD distribution whenever $\varphi$ is.

***Discrete infinite divisibility.*** A distribution on $\mathbb{N}_0$ is ID in the discrete sense if it is ID and the convolution roots also are distributions on $\mathbb{N}_0$. The left–extremity of the distribution is zero and the Lévy measure must have its mass in $\mathbb{N}$. The pgf has the form

$$P(z) = \exp\{\int_{(0,\infty)}(z^y-1)L(dy)\} = \exp\{\lambda(Q(z)-1)\} .$$

Here $\lambda$ equals the total Lévy measure and the pgf Q(z) is defined by $Q(z) = \lambda^{-1} \Sigma_{k=1}^\infty z^k L(\{k\})$. Thus a discrete ID distribution is a *compound Poisson* distribution. Setting $R(z) = \lambda Q'(z)$, we have

$$P'(z)/P(z) = R(z) = \Sigma_{k=0}^{\infty} r_k z^k ,$$

and hence

(2.3.5) $$(j+1)p_{j+1} = \Sigma_{k=0}^{j} p_{j-k} r_k , \quad j \in \mathbb{N}_0.$$

Discrete ID is equivalent to absolute monotonicity (AM) of $R(z)$ or nonnegativity of the $r_k$'s. A discrete distribution which is not ID can most often be proved to be non–ID by numerical computations.

For the NB($\beta$,q)–distribution we have $\lambda = \beta \cdot \log(p^{-1})$ and

$$Q(z) = \frac{1}{\log(p^{-1})} \Sigma_{k=1}^{\infty} \frac{q^k}{k} z^k$$

corresponding to the *Logarithmic series distribution* as found by Quenouille (1949).

***Discrete self–decomposability.*** The concept of discrete SD was introduced by Steutel & van Harn (1979). In the discrete case the operation $X \to cX$ makes no sense and is therefore replaced by *binomial thinning.* A discrete rv $X$ is SD if, for all $\pi$, $0 \le \pi \le 1$,

$$X \stackrel{d}{=} \Sigma_{i=1}^{X} Z_i + \epsilon_\pi ,$$

where all the rv's on the right hand side are independent and $Z_i \sim \mathrm{Bin}(1,\pi)$. Discrete SD implies ID and unimodality. Discrete SD holds iff $((z-1)\frac{P'(z)}{P(z)})'$ is AM or, equivalently, iff the sequence $\{r_k\}_0^\infty$ is decreasing. The NB($\beta$, q)–distribution is SD since, for this distribution, $r_k = \beta \cdot q^{k+1}$, $k \in \mathbb{N}_0$. This r–sequence is even CM. It should be added that a Poisson($\lambda$)–mixture is SD if $\lambda$ is SD.

## 2.4 Pick Functions and Some Applications

Following Donoghue (1974), we call a function $\psi(s)$ which is analytic in the upper complex half–plane ($\Im m\, s > 0$) and has nonnegative imaginary part there, a Pick function. (G. Pick was a mathematician.) We write $\psi \in \mathscr{P}$. Moreover, we write $\psi \in \mathscr{P}(a,b)$ if $\psi$ is continuous up to and on the real interval $(a,b)$ and is real there. In this case the function $\tilde{\psi}(s) = \overline{\psi(\bar{s})}$, defined for $\Im m\, s < 0$, is the unique analytic continuation of $\psi$. For a non–constant Pick function, $\Im m[\psi(s)]$ is strictly positive for $\Im m\, s > 0$, as a power series expansion of $\psi(s)$ at a possible zero of $\Im m[\psi(s)]$ easily shows. Examples of functions in the class $\mathscr{P}$ are:

a) $\psi(s) = \log(a+s) \in \mathcal{P}(-a, \infty)$
b) $\psi(s) = (s+a)^{\alpha} \in \mathcal{P}(-a, \infty),\ 0 \le \alpha \le 1,$
c) $\psi(s) = (a-s)^{-\alpha} \in \mathcal{P}(-\infty, a),\ 0 \le \alpha \le 1.$

The class $\mathcal{P}$ has the obvious but important property of being closed with respect to composition, i.e., $\psi(s) = \psi_1(\psi_2(s))$ belongs to $\mathcal{P}$ whenever $\psi_1$ and $\psi_2$ do (unless $\psi_2$ is a real constant). There is a canonical representation of a Pick function:

**Theorem 2.4.1.** *A function $\psi$ belongs to $\mathcal{P}(a, b)$, where $(a, b)$ may be empty, iff it can be represented as*

$$\psi(s) = \sigma s + \tau + \int_{-\infty}^{\infty} \left(\frac{1}{t-s} - \frac{t}{1+t^2}\right)\mu(dt)\,, \tag{2.4.1}$$

*where $\sigma \ge 0$, $\tau \in \mathbb{R}$, and $\mu$ is a nonnegative measure on $\mathbb{R}\setminus(a,b)$ satisfying $\int (1+t^2)^{-1}\mu(dt) < \infty$. Furthermore, $\sigma$, $\tau$, and $\mu$ are uniquely determined. In fact, $\tau = \Re e[\psi(i)]$, $\sigma = \lim_{\eta\uparrow\infty} \psi(i\eta)/(i\eta)$ (or $\lim_{s\to-\infty}\psi(s)/s$ if $\psi \in \mathcal{P}(-\infty, 0)$) and*

$$\int_{(c,d)} \mu(dt) = \lim_{\eta\downarrow 0} \frac{1}{\pi} \int_{(c,d)} \Im m[\psi(x+i\eta)]\,dx \qquad (i = \sqrt{-1}). \tag{2.4.2}$$

A proof is given because of the importance of this representation. The "if" part is trivial. The "only if" part is essentially a consequence of Cauchy's integral formula.

**Proof.** Consider first an analytic function $f(z)$ in $|z| < 1$; the unit disc can be mapped bijectively onto the upper half–plane. By Cauchy's integral formula,

$$f(z_0) = \frac{1}{2\pi i} \int_{|z|=r} \frac{f(z)}{z-z_0}\,dz\,, \quad |z_0| < r < 1, \tag{2.4.3}$$

where the integral is taken in the counter–clockwise direction. It follows that

$$\overline{f(0)} = \frac{1}{2\pi i} \int_{|z|=r} \frac{1}{z-z_0}\,\overline{f(z)}\,dz\,. \tag{2.4.4}$$

In fact, as some careful calculation shows, the substitution $z' = r^2/z$ transforms the right hand side of (2.4.4) into

$$\frac{1}{2\pi i} \int_{|z'|=r} \frac{1}{z'}\,h(z')\,dz'\,,$$

where $h(z') = \overline{f(\overline{z'})\cdot r^2/(r^2-z_0 z')}$. As $h$ is analytic for $|z'| < 1$ and equals $\overline{f(0)}$ at $z' = 0$, (2.4.4) follows. Subtracting then (2.4.4) from (2.4.3), we get a formula for

$f(z_0) - \overline{f(0)}$ and also, by putting $z_0 = 0$ and dividing by 2, a formula for $(f(0) - \overline{f(0)})/2 = i\Im m[f(0)]$. Subtracting the latter formula from the first one, we get

$$f(z_0) - \Re e[f(0)] = \frac{1}{2\pi i} \int_{|z|=r} \left(\frac{1}{z-z_0} - \frac{1}{2z}\right)(f(z) - \overline{f(z)})\, dz$$

$$= \frac{1}{2\pi} \int_{|z|=r} \frac{z+z_0}{z-z_0} \frac{\Im m[f(z)]}{z}\, dz = \frac{i}{2\pi} \int_0^{2\pi} \frac{re^{i\theta}+z_0}{re^{i\theta}-z_0} \Im m[f(re^{i\theta})]\, d\theta .$$

In particular,

$$\Im m[f(0)] = \int_0^{2\pi} v_r(\theta) d\theta , \quad \text{where} \quad v_r(\theta) = \frac{1}{2\pi} \cdot \Im m[f(re^{i\theta})] .$$

Assuming from now on that $\Im m[f(z)]$ is nonnegative for $|z| < 1$, we see that $\{v_r(\theta)d\theta,\ r < 1\}$ is a family of bounded nonnegative measures on the unit circle. Since the circle is compact, there is a nonnegative measure $\nu(d\theta)$ such that for each continuous, possibly complex–valued, function $g(\theta)$ on the unit circle $\int g(\theta)v_r(\theta)d\theta \to \int g(\theta)\nu(d\theta)$ as $r \to 1$ through some sequence of r–values. Choosing $g(\theta) = (e^{i\theta}+z_0)/(e^{i\theta}-z_0)$ and using that $(re^{i\theta}+z_0)/(re^{i\theta}-z_0)$ tends uniformly to $g(\theta)$ as $r \to 1$, we then get

$$f(z_0) = \Re e[f(0)] + i \int_0^{2\pi} \frac{z+z_0}{z-z_0} \nu(d\theta) , \qquad z = re^{i\theta},$$

which is the general representation formula for functions for which $\Im m[f(z)] \geq 0$ in $|z| < 1$. It is due to Herglotz (1911).

Now we let $\psi(s)$ be analytic in $\Im m\, s > 0$ with a nonnegative imaginary part there. The mapping $z \to i \cdot \frac{1-z}{1+z} = s$ maps $\{z;\ |z| \leq 1\}$ bijectively onto $\{s;\ \Im m\, s \geq 0\}$. The unit circle is mapped onto $\mathbb{R} \cup \{\infty\}$. The point $z = -1$ corresponds to $s = \infty$. The inverse mapping is $z = \frac{i-s}{i+s}$. Let $f(z) = \psi(s) = \psi(i \cdot \frac{1-z}{1+z})$. Letting now $s$ correspond to $z_0$ and $t \in \mathbb{R}$ to $z$, $|z| = 1$, we have

$$(2.4.5) \quad \psi(s) = f(z_0) = \Re e[f(0)] + i \int_0^{2\pi} \frac{z+z_0}{z-z_0} \nu(d\theta) = \Re e[f(0)] + s\tilde{\mu}(\{\infty\}) + \int_{-\infty}^{\infty} \frac{ts+1}{t-s} \tilde{\mu}(dt),$$

where $\tilde{\mu}$ is the measure induced on $\mathbb{R} \cup \{\infty\}$ by $\nu$ and the mapping. Observing that $\frac{ts+1}{t-s} = (\frac{1}{t-s} - \frac{t}{1+t^2})(1+t^2)$ and setting $\mu(dt) = (1+t^2)\tilde{\mu}(dt)$, $\sigma = \tilde{\mu}(\{\infty\})$, and $\tau = \Re e[f(0)]$, we see that (2.4.1) follows. The "only if" part is proved.

As to the uniqueness of the representation, certainly $\tau = \Re e[f(0)] = \Re e[\psi(i)]$ and, by (2.4.1) or (2.4.5), $\sigma = \lim_{\eta\uparrow\infty}\psi(i\eta)/(i\eta)$. Since $\Im m[\frac{1}{t-x-i\eta}] = \frac{\eta}{(t-x)^2+\eta^2}$, we easily get from the representation formula and by an application of Fubini's theorem

$$\lim_{\eta\downarrow 0} \frac{1}{\pi} \int_{(c,d)} \Im m[\psi(x+i\eta)]\,dx \;=\; \lim_{\eta\downarrow 0} \int_{-\infty}^{\infty} \frac{1}{\pi}\left(\arctan(\tfrac{d-t}{\eta}) - \arctan(\tfrac{c-t}{\eta})\right)\mu(dt).$$

The last integrand tends to 0 for $t \notin [c, d]$ and to 1 for $t \in (c, d)$. Since moreover the integrand multiplied by $1+t^2$ can be verified to be a bounded function with a bound not depending on $\eta$ for $\eta \leq 1$, the dominated convergence theorem shows that the limit of the last integral is $\int_{(c,d)} \mu(dt)$ provided that $\mu(\{c\}) = \mu(\{d\}) = 0$. Thus (2.4.2) holds. Of course, if $\Im m[\psi(s)]$, $\Im m\, s > 0$, is continuous up to and on the real interval (c,d) (i.e., if limit values on (c,d) exist and form a continuous function), then $\mu$ has a density there given by $\pi^{-1}\cdot\Im m[\psi(t)]$. Hence $\mu$ has no mass in (a,b) if $\psi \in \mathcal{P}(a,b)$. □

As the following examples show, it is often advantageous to modify slightly the representation (2.4.1).

**Example 2.4.1** ***(Logarithmic function).*** Consider $\psi(s) = \log s \in \mathcal{P}(0, \infty)$. Note that $\log s = \log|s| + i\cdot\arg[s]$ (with $-\pi < \arg[s] < \pi$ for $s \in \mathbb{C}\backslash(-\infty,0]$ ). Now $\Im m[\log t]$ equals $\pi$ for $t < 0$ and 0 for $t > 0$. It is assumed here and in the sequel that a real t means a point on the upper side of the real line; the negative real line is a *cut.* Hence $\mu(dt) = dt$, $t < 0$. There is no mass for $\mu$ at $t = 0$ since

$$\log s \;=\; \int_{(-\infty,0)} \left(\frac{1}{t-s} - \frac{1}{t-1}\right)dt \;=\; \int_{(-\infty,0)} \left(\frac{1}{t-s} - \frac{t}{1+t^2}\right)dt$$

as is easy to verify. The first integral representation is more natural than the second. □

**Example 2.4.2** ***(Power function).*** Let $\psi(s) = 1/(-s)^{\alpha} \in \mathcal{P}(-\infty, 0)$, $0 < \alpha < 1$. Note that $\psi(s) = \exp\{-\alpha\log(-s)\}$. Now $\Im m[\psi(t)]$ equals $\sin(\alpha\pi)\cdot t^{-\alpha}$ for $t > 0$ and 0 for $t < 0$. We also have

$$\text{(2.4.6)} \qquad \psi(s) \;=\; \int_0^{\infty} \frac{1}{t-s}\,\frac{\sin(\alpha\pi)}{\pi}\,t^{-\alpha}\,dt\,.$$

By the analyticity it suffices to verify that this representation holds for $s < 0$, and that can be done by transforming the integral into an integral of a Beta function. The representation follows also from an application of Cauchy's integral formula to $\psi(s)$ with a contour of the form

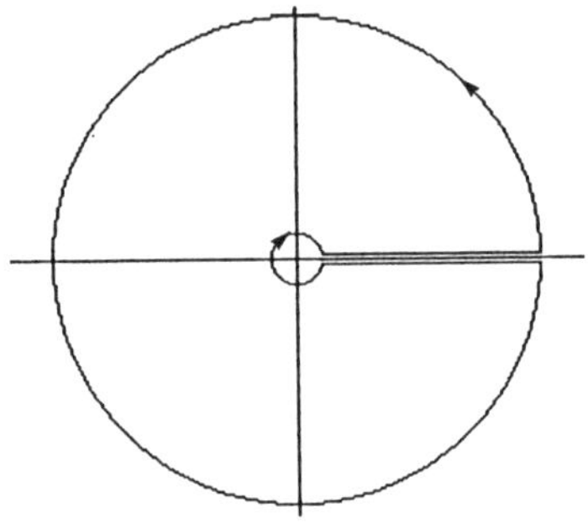

Figure 2.4.1

This is a simple general technique for verifying that a function $\psi$ belongs to $\mathscr{P}(-\infty, 0)$. It must be checked that the integrals along the circles tend to zero as the radii tend to 0 and $\infty$, respectively. □

We shall mainly be concerned with functions in $\mathscr{P}(-\infty, 0)$. Obviously these functions are increasing on $(-\infty, 0)$. Some other important simple results are now given for this class.

**Lemma 2.4.1.** *Let $\psi \in \mathscr{P}(-\infty, 0)$. If $\psi(s)$ is bounded from below as $s \to -\infty$, $s < 0$, then $\sigma = 0$ and $\int_{[1,\infty)} t^{-1}\mu(dt) < \infty$ (or, equivalently, $\int_{[0,\infty)} \frac{t}{1+t^2}\,\mu(dt) < \infty$).*

**Lemma 2.4.2.** *Let $\psi \in \mathscr{P}(-\infty, 0)$. If $\lim_{s\uparrow 0} \psi(s)$ is finite, then $\mu(\{0\}) = 0$ and $\int_{(0,1]} t^{-1}\mu(dt) < \infty$.*

**Proof of Lemmata 2.4.1 and 2.4.2.** By (2.4.1),

$$\psi(s) = \sigma s + \tau + \int_{[0,\infty)} \frac{1+st}{(t-s)(1+t^2)}\,\mu(dt).$$

As $s \to -\infty$, $\psi(s)$ tends to $\sigma\cdot(-\infty) + \tau - \int_{[0,\infty)} \frac{t}{1+t^2}\,\mu(dt)$ and Lemma 2.4.1 follows. Further, $\lim_{s\uparrow 0} \psi(s) = \tau + \int_{[0,\infty)} \frac{1}{t}\frac{1}{1+t^2}\,\mu(dt)$ and Lemma 2.4.2 follows. □

**Lemma 2.4.3.** *Let $\psi$ be analytic in $\mathbb{C}\backslash[0,\infty)$. Assume that $\lim_{s\uparrow 0} \psi(s) = 0$ and that $\liminf_{s\downarrow -\infty} \psi(s)/s \geq 0$. Then $\psi(s)$ belongs to $\mathscr{P}(-\infty, 0)$ iff $\psi(s)/s$ does.*

**Proof.** Assume that $\psi \in \mathscr{P}(-\infty, 0)$. Since $\lim_{s\uparrow 0} \psi(s) = 0$, the result in Lemma 2.4.2 implies that

$$\psi(s) = \sigma s + \tau' + \int_{(0,\infty)} \left(\frac{1}{t-s} - \frac{1}{t}\right)\mu(dt).$$

Here $\tau'$ must equal 0 and hence $\psi(s)/s = \sigma + \int_{(0,\infty)} \frac{1}{t-s} \frac{\mu(dt)}{t}$ which belongs to $\mathcal{P}(-\infty, 0)$. Conversely, if $\psi(s)/s \in \mathcal{P}(-\infty, 0)$, then

$$\psi(s)/s = \sigma's + \tau' + \int_{[0,\infty)} \left(\frac{1}{t-s} - \frac{t}{1+t^2}\right)\mu'(dt).$$

Since $\psi(s)/s$ is bounded from below as $s \downarrow -\infty$, we have by Lemma 2.4.1 that $\sigma' = 0$ and $\int_{[0,\infty)} \frac{t}{1+t^2} \mu'(dt) < \infty$. Hence

$$\psi(s)/s = \tau'' + \int_{[0,\infty)} \frac{1}{t-s} \mu'(dt),$$

where $\tau'' > 0$ since $\liminf_{s\downarrow -\infty} \psi(s)/s \geq 0$. Now since the imaginary part of $\frac{s}{t-s} = \frac{1}{t-s} - \frac{1}{t}$ is nonnegative for $\Im m\, s > 0$, it follows that $\psi \in \mathcal{P}(-\infty, 0)$. □

We now turn to some simple applications of the results given. A probability distribution on $[0,\infty)$ with cdf F is said to be a *mixture of Exponential distributions* (MED) if

$$F(x) = \int_{(0,\infty]} (1 - e^{-xt})\, M(dt), \quad x > 0,$$

where $M(dt)$, the mixing measure (for the inverse of the scale parameter), is nonnegative and satisifes $\int_{(0,\infty]} M(dt) = 1$. If $M(\{\infty\}) > 0$, then the distribution has an atom at 0. For $x > 0$ there is a pdf $f(x) = \int_{(0,\infty)} te^{-xt} M(dt)$. By Bernstein's theorem, any completely monotone pdf is an MED. The mgf of an MED equals

$$\varphi(s) = M(\{\infty\}) + \int_{(0,\infty)} \frac{1}{t-s} tM(dt),$$

which is recognized as a function in $\mathcal{P}(-\infty, 0)$. There is also a converse result.

**Theorem 2.4.2 (*MED–characterization theorem*).** *A function $\varphi(s)$ is the mgf of an MED iff $\varphi \in \mathcal{P}(-\infty, 0)$ and $\lim_{s\uparrow 0} \varphi(s) = 1$ and $\lim_{s\downarrow -\infty} \varphi(s) \geq 0$. In particular, an mgf $\varphi$ (of a distribution on $\mathbb{R}_+$) belongs to $\mathcal{P}(-\infty, 0)$ iff it corresponds to an MED.*

**Proof.** The necessity is obvious. To prove the sufficiency, note that, by Lemmata 2.4.1 and 2.4.2 we have, for some nonnegative measure $\mu(dt)$,

$$\varphi(s) = \tau' + \int_{(0,\infty)} \frac{t}{t-s} \frac{\mu(dt)}{t}.$$

Certainly $\tau' = \lim_{s\downarrow-\infty} \varphi(s) \geq 0$ and $\tau' + \int_{(0,\infty)} \frac{\mu(dt)}{t} = \lim_{s\uparrow 0} \varphi(s) = 1$. The required mixing measure is then obtained by setting $M(dt) = \mu(dt)/t$ and $M(\{\infty\}) = \tau'$. □

A simple consequence of Theorem 2.4.2 is the following result due to Goldie (1967) and Steutel (1967, 1970); cf. Feller (1971, p. 452).

**Theorem 2.4.3.** *An MED is infinitely divisible (ID) and its convolution roots are MED's.*

**Proof.** Let $\varphi$ be the mgf of an MED. Since $z^\alpha$, $0 < \alpha \leq 1$, is a Pick function, so is $\varphi_n(s) = (\varphi(s))^{1/n}$, $n \in \mathbb{N}$. An application of the "if" part of Theorem 2.4.2 then completes the proof. □

Theorem 2.4.3 shows that the Pareto and Weibull distributions with densities $f(x) = C\cdot(1+x)^{-\gamma}$ $(\gamma > 1)$ and $f(x) = C\cdot x^{\alpha-1}\exp\{-x^\alpha\}$ $(0 < \alpha \leq 1)$, respectively, are ID.

From the fact that $\log z$ is a Pick function it follows that there is an exponential representation of the mgf of an MED; we return to that in Chapter 9.

Any *mixture of Geometric distributions* can be obtained as the distribution of $X(T)$ where $X(\tau)$ is a Poisson process and $T \sim$ MED; hence such a mixture is (discrete) ID. Can the probability generating function (pgf) $P(z)$ of such a mixture also be characterized by the Pick function property? The answer is positive but there is a complication worth noticing. We have with the same proof as for Theorem 2.4.2:

**Theorem 2.4.4.** *A function $P(z)$ is the pgf of a mixture of Geometric distributions iff $P(z) \in \mathcal{P}(-\infty, 1)$ and*

$$(a)\ \lim_{z\uparrow 1} P(z) = 1; \qquad (b)\ \lim_{z\downarrow-\infty} P(z) \geq 0\,.$$

*If $P(z)$ is assumed to be a pgf, then (a) can be deleted but not (b).*

In fact, the pgf of a mixture of a Geometric distribution and a distribution with all its mass at the point 1, is a Pick function. That mixture is not a mixture of Geometric distributions.

## 2.5 Some Complementary Results

The results briefly presented in this section will be little used in the sequel but belong to the more or less necessary background. They complement Steutel's (1979) survey on ID.

I (***Properties of the MED–class***). Some simple properties of the MED–class are worth noticing.

a) If $X \sim \text{MED}$ and $Y \geq 0$ are independent rv's, then $X \cdot Y \sim \text{MED}$; this is merely a reformulation of the fact that the MED–class is preserved under mixing.

b) If $X \sim \text{MED}$, then $X^q \sim \text{MED}$ for $q \geq 1$. This follows from Bernstein's theorem since $\Pr(\{X^q > x\}) = \overline{F}_X(x^{1/q})$ is completely monotone (CM), $x^{1/q}$ has a CM derivative. Analogously, since $\log(x+1)$ has a CM derivative, $e^X - 1 \sim \text{MED}$ when $X \sim \text{MED}$.

c) If $f_1$ and $f_2$ are densities of MED's on $(0,\infty)$, then so is $C \cdot f_1 \cdot f_2$, where $C$ is a normalizing constant. (It is assumed that $f_1 \cdot f_2$ is integrable.)

II (***Logconvexity and infinite divisibility***). A logconvex pdf $f$ on $(0,\infty)$ is ID; Steutel (1970). (A possible atom at $0$ does not change the conclusion.) The class, LC, of logconvex densities, which contains the CM densities, is preserved under mixing (by Artin's theorem), multiplication, and formation of powers with positive exponents. If $X \sim \text{LC}$, then $X^q \sim \text{LC}$ for $q \geq 1$ and $e^X - 1 \sim \text{LC}$. If moreover $Y > 0$ is independent of $X$, then $X \cdot Y \sim \text{LC}$. It can be shown that LC is not preserved under convolution roots.

III (***Infinitely divisible distributions on*** $\mathbb{R}$). A symmetric pdf $f$ on $\mathbb{R}$ such that $f(x)$ is CM for $x > 0$ is ID; Steutel (1968, 1970). In fact, $f$ can be shown to be the pdf of $X(T)$, where $X(\tau) \sim N(0,\tau)$ and $T \sim \text{MED}$. It seems to be unknown whether logconvexity for $f(x)$, $x > 0$, guarantees ID also.

IV (***More infinite divisibility***). If $f$ is a positive decreasing pdf on $(0,\infty)$ such that, for all $x > 0$,

$$\frac{f'(y)}{f(y)} \leq \frac{1}{x} + \frac{f'(x)}{f(x)}, \quad 0 < y \leq x,$$

then $f$ is ID. The class of such densities and limits thereof contains the logconvex densities (for which $f'/f \uparrow$). It is not preserved under mixing but has otherwise properties which are similar to those of the class LC.

V (***Discrete ID***). If $(p_j)_0^\infty$ is a decreasing probability function (pf) on $\mathbb{N}_0$ such that

$$\max_{0\le k\le j} \frac{p_{k+1}}{p_k} \le \frac{j+1}{j}\cdot\frac{p_{j+1}}{p_j}, \quad j = 1, 2, 3, \ldots ,$$

then it is ID; Katti (1979). In particular, a logconvex pf (for which $p_{j+1}/p_j \uparrow$ ) is ID. Chang (1989) has given another sufficient condition for discrete ID.

VI (***Self–decomposability***). A positive strictly decreasing pdf f on $(0,\infty)$ satisfying

$$\frac{f'(y)}{f(y)} \le \frac{1}{x} + \frac{f''(x)}{f'(x)}, \quad 0 < y \le x,$$

is self–decomposable (SD). In particular, SD holds for the class $\mathcal{A}$ (say) of logconvex pdf's of the form $f(x) = f(1)\cdot\exp\{-\int_1^x g(y)dy\}$, where g is a positive decreasing function such that $yg(y)$ is increasing. *We have* $X \sim \mathcal{A}$ *iff* $X$ *has a logconvex pdf and* $\log X$ *has a logconcave pdf.* The class $\mathcal{A}$ is closed with respect to multiplication of densities and with respect to multiplication of independent rv's. Moreover, if $X \sim \mathcal{A}$, then $X^q \sim \mathcal{A}$ for $q \ge 1$. Every pdf which is proportional to the Laplace transform of an SD distribution belongs to $\mathcal{A}$.

VII (***Discrete SD***). A strictly decreasing pf $(p_j)_0^\infty$ on $\mathbb{N}_0$ such that

$$\max_{0\le k\le j} \frac{p_{k+1}}{p_k} \le \frac{j+2}{j+1}\cdot\frac{p_{j+1} - p_{j+2}}{p_j - p_{j+1}}, \quad j = 0, 1, 2, \ldots ,$$

is SD in the discrete sense. In particular a logconvex pf of the form

$$p_j = p_0\cdot\prod_{k=1}^{j}(1-\rho_k), \quad j = 1, 2, 3, \ldots ,$$

where $\rho_k \downarrow$ and $k\cdot\rho_k \uparrow$, is SD in the discrete sense.

Proofs of the statements in IV – VII can be found in B1987, i.e. Bondesson (1987). The integral equation (2.3.4) and its discrete analogue (2.3.5) play a fundamental role in the proofs.

# 3. GENERALIZED GAMMA CONVOLUTIONS

This chapter and Chapters 4–7, which treat generalized Gamma convolutions (GGC), form the central body of this work. The class of GGC's was introduced by O. Thorin in 1977 as a useful tool for proving ID of particular distributions. This class of distributions, in honor of Thorin often called the $\mathcal{T}$–class, also has its own interest. It is the smallest class of distributions on $\mathbb{R}_+$ that contains the Gamma distributions and is closed with respect to convolution and weak limits. The $\mathcal{T}$–class is surprisingly rich, and it has a remarkable structure as will be seen. It should be added that Thorin's original name for a GGC was "generalized Γ–convolution" which is more clear but somewhat hard to abbreviate.

The chapter starts with the formal definition of a GGC and the basic theory. Complex analysis is used. In Section 3.2 several examples of GGC's are given. The remaining sections treat more special topics.

## 3.1 Definition and Basic Theory

The mgf of a finite convolution of Gamma distributions, Gamma($u_j$,$t_j$), j = 1, ..., n, has the form

$$\varphi(s) = \prod_{j=1}^{n} \left(\frac{t_j}{t_j-s}\right)^{u_j} = \exp\{\sum_{j=1}^{n} \log\left(\frac{t_j}{t_j-s}\right) u_j\}.$$

Permitting also a Gamma distribution to be degenerate at a point $a > 0$ (let $\beta \to \infty$ in Gamma($\beta$, $\beta/a$)), we may add a factor $e^{as}$ on the right hand side. Considering now also weak (non–defective) limits of such convolutions, i.e., pointwise limits of such mgf's, we are led to the following definition.

**Definition.** *A generalized Gamma convolution (or $\mathcal{T}$–distribution) is a probability distribution $F$ on $\mathbb{R}_+ = [0,\infty)$ with mgf of the form*

$$\text{(3.1.1)} \quad \varphi(s) = \int e^{sx}F(dx) = exp\{as + \int log\left(\frac{t}{t-s}\right)U(dt)\}, \quad s \leq 0 \quad (or\ s \in \mathbb{C}\setminus(0,\infty)),$$

*where $a \geq 0$ and $U(dt)$ is a nonnegative measure on $(0,\infty)$ satisfying*

$$\text{(3.1.2)} \qquad \int_{(0,1]} |\log t|\ U(dt) < \infty \quad and \quad \int_{(1,\infty)} t^{-1}U(dt) < \infty .$$

**Remark.** Unless otherwise indicated, in the sequel all integrals are calculated over their natural domains, most often $(0, \infty)$. □

The conditions (3.1.2) are necessary and sufficient for the convergence of the integral in (3.1.1). Certainly (3.1.1) holds iff $\varphi(0) = 1$ and

$$\varphi'(s)/\varphi(s) = a + \int \frac{1}{t-s}\, U(dt)\,. \tag{3.1.3}$$

Hence, for a GGC, $\varphi'(s)/\varphi(s) \in \mathcal{P}(-\infty, 0)$ and the Thorin measure U equals the canonical measure $\mu$ for Pick functions. Thus $a$ and U are uniquely determined. In fact, $a$ is the left–extremity of the distribution. For an rv $X \sim \mathcal{T}$, we have

$$E[X] = \varphi'(0) = a + \int t^{-1}U(dt) \quad \text{and} \quad Var[X] = \varphi''(0)\varphi(0) - (\varphi'(0))^2 = \int t^{-2}U(dt).$$

Obviously $\mathcal{T}$ is closed with respect to translation to the right and change of scale of its members and also closed with respect to convolution and convolution roots. By the *Closure Theorem*, to be proved only at the end of this section, it is further closed with respect to weak (non–defective) limits. Moreover, it precisely equals the class of weak limits of finite convolutions of Gamma distributions.

Since the Gamma distribution is ID and self–decomposable (SD), so is every GGC. There is a simple characterization of the GGC's in terms of the Lévy measure.

**Theorem 3.1.1.** *A probability distribution on $\mathbb{R}_+$ is a GGC iff it is ID and the Lévy measure has a density $\ell$ such that $y\ell(y)$, $y > 0$, is completely monotone. In fact, $y\ell(y) = \int e^{-yt}U(dt)$.*

**Proof.** Let $\varphi$ be the mgf of a GGC. Then

$$\frac{\varphi'(s)}{\varphi(s)} = a + \int \frac{1}{t-s}\, U(dt) = a + \int \left(\int e^{-y(t-s)}dy\right) U(dt) = a + \int e^{sy}\left(\int e^{-yt}U(dt)\right)dy.$$

Hence $y\ell(y) = \int e^{-yt}U(dt)$, cf. (2.3.3), and thus $y\ell(y)$ is CM. Conversely, if $y\ell(y)$ is CM and thus equals the LT of a nonnegative measure U on $\mathbb{R}_+$, then this measure has no atom at 0 since $\ell$ is integrable over $(1, \infty)$. □

Since any non–degenerate SD distribution is absolutely continuous and unimodal, so is any non–degenerate GGC. The pdf is strictly positive for $x > a$; cf. Section 4.5. For a non–degenerate GGC with left–extremity 0 and pdf f, the total U–measure is given by

$$\int U(dt) = \sup\{\alpha\,;\, \lim_{x \downarrow 0} f(x)/x^{\alpha-1} = 0\} \tag{3.1.4}$$

This relation will be proved only in Chapter 4.

There is also an important interpretation of the measure U and its total in terms of the mgf $\varphi$. Obviously $\varphi$ is a well–defined analytic function in $\mathbb{C}\backslash[0,\infty)$, the complex plane cut along the nonnegative real axis. The mgf has no zeros there and takes conjugate values at conjugate points. If, moreover, U has a density being for instance continuous on $(0,\infty)$, then $\varphi$ is also continuous up to and on $(0,\infty)$ but, in general, with different values on the two sides of the cut $[0,\infty)$. Since $\log z = \log|z| + i\cdot\arg[z]$, we then have for $s > 0$ *on the upper side of the cut*

$$\varphi(s) = \exp\{as + \int \left(\log\left|\frac{t}{t-s}\right| + i\pi\cdot H(s-t)\right) U(dt)\},$$

where $H(y) = 1$ for $y \geq 0$ and $0$ elsewhere. (On the lower side, $i\pi$ is changed to $-i\pi$.) In the sequel, an $s > 0$ will always mean a point on the upper side of the cut unless otherwise stated. It follows that

$$\frac{1}{\pi}\cdot\arg[\varphi(s)] = \int_{(0,s]} U(dt)\ , \quad s > 0. \tag{3.1.5}$$

Hence the argument of the mgf is increasing when s moves from 0 to $\infty$ along the cut. The path described by $\varphi(s)$, $s \in \mathbb{R}$, goes around the origin in the positive direction as s moves from $-\infty$ to $\infty$. See Figure 3.1.1. Often, but not always, $\varphi(s) \to 0$ as $s \to \infty$; this holds if $a = 0$ and $\lim_{s\to\infty} \int \log\left|\frac{t}{t-s}\right| U(dt) = -\infty$.

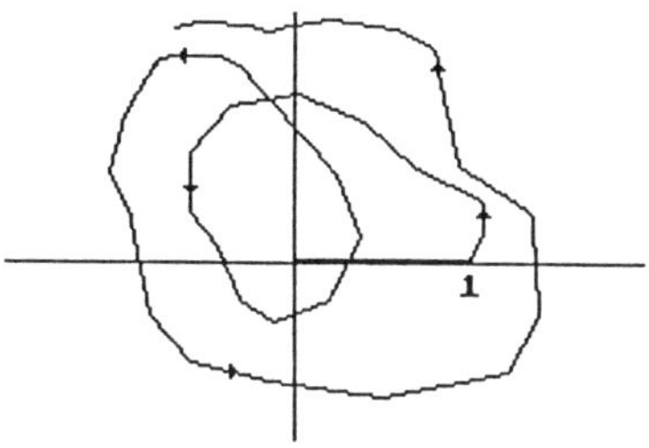


Figure 3.1.1. *Path of $\varphi(s)$, $s \in \mathbb{R}$.*

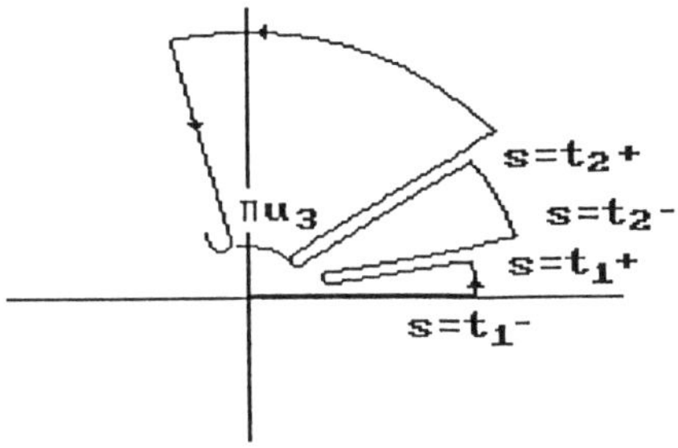


Figure 3.1.2. *Path of $\varphi(s)$ for U discrete.*

In particular, $\pi\cdot\int U(dt)$ equals the total change of the argument of $\varphi(s)$ when s moves from $-\infty$ (or 0) to $\infty$. Thus the path goes infinitely many times around the origin iff $\int U(dt) = \infty$.

With a proper interpretation, (3.1.5) continues to hold when the measure U does not satisfy any regularity conditions. For simplicity, let us consider the case when U is a discrete measure with mass $u_j$ at $t_j$, $j = 1, \ldots, n$; $t_1 < t_2 < \ldots < t_n$. In this case $\varphi(s)$

is not defined at the points $t_j$. To avoid this difficulty, we let $s$ move around small half–circles in the upper half–plane with radii $\epsilon$ and centres at $t_j$. The path of $\varphi(s)$, $s \in \mathbb{R}$, $s \neq t_j$, has the appearance shown in Figure 3.1.2. For $\epsilon$ very small, the arcs are very far from the origin.

There is another interpretation of the total $\int U(dt)$ when it is finite. Consider $\varphi(s)$ when $s$ moves in the positive direction along a very large circle with radius $r$ and centre at $0$. At $s = r$, $\arg[\varphi(s)] \approx \pi \int U(dt)$. The argument then decreases to $0$ at $s = -r$ and to about $-\pi \int U(dt)$ at the point $r$ on the negative side of the cut. The total change in argument is $-2\pi \int U(dt)$.

We shall now look more carefully at the mgf of a GGC from the Pick function point of view. It has already been seen that for a GGC

$$\varphi'(s)/\varphi(s) = a + \int_{[0,\infty)} \frac{1}{t-s}\, U(dt) \tag{3.1.6}$$

and hence $\varphi'(s)/\varphi(s)$ belongs to $\mathscr{P}(-\infty, 0)$. In fact, this is also a characteristic property for the mgf of a GGC.

**Theorem 3.1.2** ***(Pick function characterization).*** *A probability distribution on $\mathbb{R}_+$ is a GGC iff its mgf $\varphi$ is analytic and zero–free in $\mathbb{C}\setminus[0,\infty)$ and $\Im m[\varphi'(s)/\varphi(s)] \geq 0$ for $\Im m\, s > 0$.*

**Proof.** Only the "if" part has to be verified. Let $\psi(s) = \varphi'(s)/\varphi(s) \in \mathscr{P}(-\infty, 0)$. Then $\psi$ can be represented by (2.4.1) where $\mu$ has no mass on $(-\infty, 0)$. Certainly, since $\varphi(s)$ is an mgf, $\psi(s) \geq 0$ for $s < 0$. Hence, by Lemma 2.4.1, (3.1.6) holds with $a = \tau - \int_{[0,\infty)} \frac{t}{1+t^2}\, \mu(dt)$ and $U(dt) = \mu(dt)$. It is possible that $U(\{0\}) > 0$. Letting $s \to -\infty$ in (3.1.6) and using the fact that $\psi(s) \geq 0$ for $s < 0$, we see that $a \geq 0$. Moreover, from the fact that $\varphi(0) = 1$ (or just finite) it is easy to see by an integration of (3.1.6) that $U(\{0\}) = 0$. That integration also gives (3.1.1). □

When $\varphi'/\varphi$ is continuous up to and on the cut, then by the inversion formula (2.4.2) for Pick functions

$$\int_{(c,d)} U(dt) = \frac{1}{\pi} \cdot \int_{(c,d)} \Im m[\varphi'(s)/\varphi(s)]\, ds\,. \tag{3.1.7}$$

As $\Im m[\varphi'(s)/\varphi(s)] = \Im m[\frac{d}{ds}\log\varphi(s)] = \frac{d}{ds}\arg[\varphi(s)]$ for $s > 0$, we have rederived (3.1.5).

There is a generalization of Theorem 3.1.2 making it unnecessary to check that $\varphi(s)$ is zero–free.

**Theorem 3.1.3** ***(Main characterization theorem).*** *The mgf* $\varphi$ *of a probability distribution on* $\mathbb{R}_+$ *corresponds to a GGC iff* $\varphi$ *is analytic in* $\mathbb{C}\backslash[0,\infty)$ *and*

$$\Im m[\varphi'(s)\cdot\overline{\varphi(s)}] \geq 0 \text{ for } \Im m\, s > 0\,. \tag{3.1.8}$$

**Proof.** Since $\varphi'(s)/\varphi(s) = \varphi'(s)\cdot\overline{\varphi(s)}/|\varphi(s)|^2$, Theorem 3.1.2 shows that it suffices to verify that $\varphi$ has no zeros if (3.1.8) holds. Certainly $\varphi(s) \neq 0$ for $s < 0$. Assume that $\varphi(s_0) = 0$ for some $s_0$ in $\Im m\, s > 0$. Now, for some $k \geq 1$ and some $c_k \neq 0$,

$$\varphi(s) = c_k(s-s_0)^k + O(|s-s_0|^{k+1}), \qquad \varphi'(s) = kc_k(s-s_0)^{k-1} + O(|s-s_0|^k)$$

for $s$ close to $s_0$. Hence

$$\Im m[\varphi'(s)\cdot\overline{\varphi(s)}\,] = k|c_k|^2|s-s_0|^{2k-2}\Im m[\overline{s-s_0}\,] + O(|s-s_0|^{2k})\,.$$

For $s-s_0 = i\eta$, $\eta > 0$ and $\eta$ sufficiently small, a contradiction is obtained. □

The Main Characterization Theorem is very powerful, as will be seen in Chapter 4 in particular. There are simple geometric interpretations of $J(s) = \Im m[\varphi'(s)\cdot\overline{\varphi(s)}]$, $s > 0$. In fact, as can be verified, $J(s) = 2\cdot A'(s)$, where $A(s)$ is the area swept over by (the vectors) $\varphi(t)$, $t \leq s$; cf. Fig. 3.1.1. Moreover, $J(s)$ equals the scalar product of the vector $i\varphi$ (i.e., $\varphi$ rotated $90^\circ$) and the tangent vector $\varphi'$.

Sometimes it is possible to verify that an mgf $\varphi$ corresponds to a GGC by mainly looking at the values of $\varphi(s)$ for $s > 0$.

**Theorem 3.1.4** ***(Inversion theorem).*** *Let* $\varphi$ *be the mgf of a probability distribution* $F$ on $\mathbb{R}_+$. *Assume that*

a) $\varphi$ *is analytic and zero-free in* $\mathbb{C}\backslash[0,\infty)$ *and moreover continuously differentiable up to and on the cut with non-zero boundary values.*
b) $\varphi'(s)/\varphi(s) \to a$ *uniformly as* $|s| \to \infty$, $s \in \mathbb{C}\backslash[0,\infty)$.
c) $\arg[\varphi(s)]$ *is increasing for* $s > 0$, *or, equivalently,* $\Im m[\varphi'(s)/\varphi(s)] \geq 0$ *for* $s > 0$.

*Then* $F$ *is a GGC with left-extremity* $a$ *and* $\int_{(0,s]} U(dt) = \frac{1}{\pi}\cdot\arg[\varphi(s)]$, $s > 0$. *Moreover,* $U$ *has a density given by* $u(t) = \frac{1}{\pi}\cdot\Im m[\varphi'(t)/\varphi(t)]$.

This inversion theorem easily follows by an application of Cauchy's integral formula to the function $\varphi'(s)/\varphi(s) - a$ with a contour as in Fig. 2.4.1.

**Remark 1.** The regularity conditions in (a) can be weakened. For example, $\varphi'$ may have discontinuities at $s = t_j > 0$, $j = 1, ..., n$, such that, for some $\epsilon > 0$, $\varphi'(s)/\varphi(s) = \frac{u_j}{t_j - s} + O((s-t_j)^{\epsilon-1})$ $(u_j \geq 0)$ as $s \to t_j$; the singularities can be avoided by an integration along small half–circles. The U–measure gets the mass $u_j$ at $t_j$. A singularity is harmless if $u_j = 0$. □

**Remark 2.** Zero–freeness of $\varphi$ on the cut holds if $\Im m[\varphi'(s)\cdot\overline{\varphi(s)}] > 0$, $s > 0$. By the *principle of argument*, $\varphi$ has no zeros in $\mathbb{C}\backslash[0,\infty)$ if the change of the argument of $\varphi(s)$ as $s$ moves along the large circle in Figure 2.4.1 in the limit equals $-2\cdot\arg[\varphi(+\infty)]$. □

The Inversion Theorem has played a dominant role in earlier works on GGC's. However, in this exposition the Main Characterization Theorem with only one condition to verify will be the main tool. It is appropriate to add that there is another inversion formula which gives the pdf $f(x)$ expressed in $\varphi$. We have, under some regularity conditions on $\varphi$ which are not made precise,

$$f(x) = \frac{1}{\pi}\Im m[\int_0^\infty e^{-xt}\varphi(t)dt] = \frac{1}{\pi}\int_0^\infty e^{-xt}\Im m[\varphi(t)]dt. \tag{3.1.9}$$

This formula, valid for all distributions with mgf's which are analytic in $\mathbb{C}\backslash[0,\infty)$, can be derived from the Fourier inversion theorem by a change of the path of integration; see, e.g., Watson (1981, Chapter 3). This formula is not used much here.

Finally in this section we return to a formulation and proof of the Closure Theorem. It is convenient to modify slightly the representation (3.1.1). We define a nonnegative measure $\nu(dt)$ on $[0,\infty]$ by $\nu(\{0\}) = 0$, $\nu(\{\infty\}) = a$, and $\nu(dt) = \log(1+t^{-1})U(dt)$, $0 < t < \infty$. Setting also $g(s,t) = \frac{1}{\log(1+t^{-1})}\log(\frac{t}{t-s})$, we have instead of (3.1.1): $\varphi(s) = \exp\{\int_{[0,\infty]}g(s,t)\nu(dt)\}$.

**Theorem 3.1.5** ***(Closure theorem).*** *Let $F_n \in \mathcal{T}$, $n = 1, 2, ...,$ be defined by the measures $\nu_n$, $n = 1, 2, ...$ If $F_n \to F$ weakly (with $F$ non–defective), then $F \in \mathcal{T}$ and the corresponding measure $\nu$ is a vague limit of $\nu_n$ on $[0,\infty]$. Conversely, if $\nu_n \to \nu$ vaguely on $[0,\infty]$ and $\nu(\{0\}) = 0$, then $F_n \to F$ weakly, where $F \in \mathcal{T}$ corresponds to $\nu$.*

**Proof.** Let $F_n \to F$ weakly. Then $\log\varphi_n(s) \to \log\varphi(s)$, $s < 0$. Since $F$ is non–defective $\lim_{s\uparrow 0}\varphi(s) = 1$. Now $\log\varphi_n(-1) = \int_{[0,\infty]}\nu_n(dt)$ and hence $\{\nu_n\}_1^\infty$ is a sequence of bounded measures on $[0,\infty]$. Let $\{\nu_{n_j}\}_{j=1}^\infty$ be any vaguely convergent subsequence with limit $\nu$. Now, for $s < 0$, the function $t \to g(s,t)$ is continuous on $[0,\infty]$ with value $-1$

at $t = 0$ and value $s$ at $t = \infty$. Hence

$$\log\varphi_{n_j}(s) \to \int_{[0,\infty]} g(s,t)\nu(dt) = -\nu(\{0\}) + as + \int_{(0,\infty)} \log(\tfrac{t}{t-s})\, U(dt),$$

where $a$ and $U$ are defined in the obvious way by $\nu$. Since the limit also equals $\log\varphi(s)$ which tends to $0$ as $s \uparrow 0$, certainly $\nu(\{0\}) = 0$. Hence $F \in \mathcal{T}$. Moreover, the vague limit $\nu$ is uniquely determined and hence the sequence $\{\nu_n\}$ itself is vaguely convergent.

The proof of the converse part is rather immediate and is omitted. □

The theorem can be reformulated as follows: The sequence $F_n \subset \mathcal{T}$, $n - 1, 2, \ldots$, tends weakly to $F \in \mathcal{T}$ iff

i) $U_n \to U$ vaguely on $(0,\infty)$

ii) $a = \lim_{A\to\infty} \lim_{n\to\infty} (a_n + \int_{(A,\infty)} t^{-1} U_n(dt))$

iii) $\lim_{\epsilon\to 0} \lim_{n\to\infty} \int_{(0,\epsilon)} (\log t^{-1}) U_n(dt) = 0$.

If (iii) fails to hold then the limit distribution is defective. An arbitrary GGC defined by $a$ and $U$ is certainly a limit of a sequence of finite convolutions of Gamma distributions; we choose $a_n = 0$ and let $U_n$ be discrete with only a finite number of atoms and such that (i) – (iii) hold.

## 3.2 Examples of Distributions in $\mathcal{T}$

This section is devoted to examples of GGC's for which simple analytic expressions for the mgf can be found. These examples also illustrate the basic technique and prepare for what follows. In Chapters 4 – 6 many GGC's with no explicit expressions for the mgf are presented.

**Example 3.2.1** ***(Stable distribution).*** Consider the mgf $\varphi(s) = \exp\{-(-s)^\alpha\}$, $0 < \alpha < 1$, of a (strictly) Stable distribution on $(0, \infty)$. It is known from Section 2.3 that $r(y) = y\ell(y) = \frac{\alpha}{\Gamma(1-\alpha)} \cdot y^{-\alpha}$. Thus, by Theorem 3.1.1, this distribution is a GGC since $y\ell(y)$ is CM. Further,

$$y\ell(y) = \int e^{-yt} u(t)dt\,, \quad \text{where} \quad u(t) = \frac{\alpha}{\Gamma(1-\alpha)} \cdot \frac{1}{\Gamma(\alpha)}\, t^{\alpha-1} = \frac{\alpha \sin(\alpha\pi)}{\pi}\, t^{\alpha-1}.$$

Hence $u$ is the density of the U–measure, the total of which equals $\infty$.

We can also obtain this result without using the Lévy measure. In fact, by Theorem 3.1.2, the distribution is a GGC since $\varphi'(s)/\varphi(s) = \alpha\cdot(-s)^{\alpha-1}$ is a Pick function. Moreover, the density of U equals

$$u(t) = \frac{1}{\pi}\cdot\Im m[\varphi'(t)/\varphi(t)] = \frac{\alpha}{\pi}\cdot\Im m[(-t)^{\alpha-1}]$$

$$= \frac{\alpha}{\pi}\cdot\Im m[t^{\alpha-1}\exp\{(\alpha-1)\cdot(-i\pi)\}] = \frac{\alpha}{\pi}\cdot t^{\alpha-1}\sin(\pi(1-\alpha)) = \frac{\alpha\sin(\alpha\pi)}{\pi}t^{\alpha-1}.$$

Note that $\arg[-t] = -i\pi$ since $-t$ is on the lower side of the negative real axis by the convention that $t$, $t > 0$, is a point on the upper side of the positive real axis.

For $s > 0$ we have $\varphi(s) = \exp\{-|s|^{\alpha}(\cos(\alpha\pi) - i\cdot\sin(\alpha\pi))\}$. The following figures show that the path $\mathbb{R}_+ \ni s \to \varphi(s)$ has different appearances for $\alpha < 1/2$, $\alpha = 1/2$, and $\alpha > 1/2$.

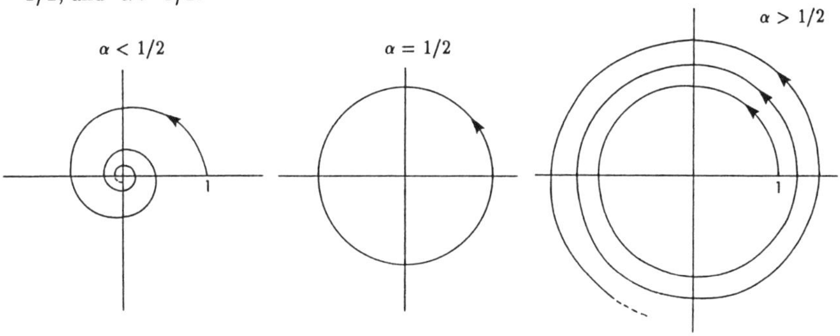


Figure 3.2.1. *Paths of $\varphi(s)$ for Stable distributions.*

**Example 3.2.2** ***(Polya densities).*** The pdf of a denumerable convolution of Exponential distributions (and possibly a degenerate distribution) is often called a Polya frequency function on $\mathbb{R}_+$ of order $\infty$. The class of such pdf's f is denoted $PF_\infty$. As shown by Schoenberg (1951), the densities in this class have the characteristic property of being totally positive of all orders, i.e., for each $n \geq 2$, every $n \times n$ matrix $\mathbf{M}$ formed by elements $m_{ij} = f(x_i - y_j)$, $x_1 < x_2 < ... < x_n$, $y_1 < y_2 < ... < y_n$, has a nonnegative determinant. (For $x_i < y_j$, $m_{ij} = 0$.) The major document on total positivity is the book by Karlin (1968). Marshall & Olkin (1979, Chapter 18) provide a brief account. Total positivity of order 2 is equivalent to logconcavity, or strong unimodality. Obviously the $PF_\infty$-class equals the subclass of $\mathcal{T}$ for which the U-measure is discrete with atoms with integral mass. The closure theorem for GGC's guarantees in particular that also the $PF_\infty$-class is closed with respect to weak limits; note that atoms with mass 1 cannot be split in the limit.

In non–parametric and multivariate statistics many important distributions correspond to weighted sums $\Sigma c_i X_i^2$, where $c_i > 0$ and the rv's $X_i$ are independent and $N(0,1)$–distributed. Since $X_i^2 \sim \text{Gamma}(1/2,1/2)$, these limit distributions are also GGC's. The U–measure is discrete and all atoms have mass $1/2$. □

**Example 3.2.3** ***(First passage time distribution).*** Let a particle perform a random walk in continuous time on the lattice of integers $\mathbb{Z}$. It moves at the epochs in a Poisson process of unit intensity. With probability $p$ $(p \geq 1/2)$, the particle moves one step to the right and otherwise one step to the left. We are interested in the random time $T$ it takes to reach the point $1$ when the start is at the point $0$.

As is easily verified, the mgf $\varphi$ of $T$ satisfies the equation

$$\varphi(s) = p\cdot\frac{1}{1-s} + q\cdot\frac{1}{1-s}(\varphi(s))^2 \qquad (q = 1-p).$$

There are two solutions but only one is an mgf:

$$\varphi(s) = \frac{1-s - \sqrt{(1-s)^2 - 4pq}}{2q}. \tag{3.2.1}$$

Let $\alpha = 2\sqrt{pq} \leq 1$. By differentiation,

$$\varphi'(s)/\varphi(s) = \frac{1}{\sqrt{(1-s)^2 - \alpha^2}} = (1-\alpha-s)^{-1/2}(1+\alpha-s)^{-1/2}.$$

By (3.2.1), $\varphi(s)$ is analytic and zero–free in $\mathbb{C}\backslash[1-\alpha, 1+\alpha]$. Now, for $t \in (1-\alpha, 1+\alpha)$ (on the upper side), we have

$$\frac{1}{\pi}\cdot\Im\text{m}[\varphi'(t)/\varphi(t)] = \frac{1}{\pi}\cdot(t-1+\alpha)^{-1/2}(1+\alpha-t)^{-1/2}$$

since $\Im\text{m}[\exp\{-(-i\pi)/2\}] = 1$. By Theorem 3.1.4 and its Remark 1, it follows that the passage time distribution is a GGC with $a = 0$ and U–measure having the Beta$(1/2,1/2)$–density

$$u(t) = \frac{1}{\pi}\cdot(t-1+\alpha)^{-1/2}(1+\alpha-t)^{-1/2}, \qquad 1-\alpha < t < 1+\alpha.$$

We have assumed above that $p \geq 1/2$ ensuring that $T$ is finite with probability 1. For $p < 1/2$, the only difference is that

$$\varphi(s) = \varphi(0)\cdot\exp\{\int \log(\tfrac{t}{t-s})\, u(t)dt\},$$

where now $\varphi(0) = p/q < 1$; the GGC is defective.

The total mass of U happens to be 1. It follows that the distribution is also a mixture of Exponential distributions. This is true for every GGC with $a=0$ and $\int U(dt) \le 1$. In fact, since $\log(\frac{t}{t-s})$ is a Pick function, the imaginary part of $\int \log(\frac{t}{t-s})U(dt)$ is positive for $\Im m\, s > 0$ and less than $\pi$ and hence $0 < \arg[\varphi(s)] < \pi$, i.e., $\Im m[\varphi(s)] > 0$. By Theorem 2.4.2 the assertion follows. □

**Example 3.2.4** ***(Products of rv's).*** Consider the product $X = Y^{1/\alpha}\cdot X_\alpha$, where Y and $X_\alpha$ are independent rv's and $Y \sim \text{Gamma}(\gamma, 1)$ and $X_\alpha$ is (strictly) Stable with index $\alpha < 1$ and unit scale. The mgf of X is $\varphi(s) = (1+(-s)^\alpha)^{-\gamma}$. The distribution of X is self–decomposable; cf. Section 2.3. It is even a GGC as we shall see now. Certainly $\varphi(s)$ is analytic and zero–free in $\mathbb{C}\backslash[0,\infty)$. Now

$$\varphi'(s)/\varphi(s) = \frac{\gamma\alpha\cdot(-s)^{\alpha-1}}{1+(-s)^\alpha},$$

and hence, for $\Im m\, s > 0$,

$$\begin{aligned}\Im m[\varphi'(s)/\varphi(s)] &= \gamma\alpha\cdot\Im m[(-s)^{\alpha-1}\cdot(1+(-\bar{s})^\alpha)]\cdot|1+(-s)^\alpha|^{-2} \\ &= \gamma\alpha\cdot\left(\Im m[(-s)^{\alpha-1}] + |s|^{2\alpha}\,\Im m[-s^{-1}]\right)\cdot|1+(-s)^\alpha|^{-2} > 0\end{aligned}$$

since $(-s)^{\alpha-1}$ and $-s^{-1}$ are Pick functions. By Theorem 3.1.2, $X \sim \mathcal{T}$. Further, the U–measure has the density

$$u(t) = \frac{1}{\pi}\cdot\Im m[\varphi'(t)/\varphi(t)] = \frac{1}{\pi}\cdot\frac{\gamma\alpha\cdot t^{\alpha-1}\sin(\alpha\pi)}{1+t^{2\alpha}+2t^\alpha\cos(\alpha\pi)}, \quad t > 0,$$

as is easily verified; cf. Example 3.2.1.

The result here is a special case of more general results for products of rv's given in Chapters 5 and 6 and also a special case of a result in Section 3.3 which concerns composition of mgf's. □

**Example 3.2.5** ***(Exponential family).*** Let f(x) be the pdf of a GGC with mgf $\varphi(s)$. Consider for $\theta > 0$ the new pdf $f(x;\theta) = C\cdot e^{-\theta x}f(x)$, where $C = 1/\varphi(-\theta)$. It has the mgf $\varphi(s;\theta) = \varphi(s-\theta)/\varphi(-\theta)$. Hence

$$\varphi'(s;\theta)/\varphi(s;\theta) = a + \int \frac{1}{t+\theta-s}\,U(dt)$$

and it is seen that $f(x;\theta)$ is the pdf of a GGC with U–measure $U(dt;\theta) = U(dt+\theta)$; i.e., the original U is shifted $\theta$ units to the right. We have an exponential family of GGC's. □

**Example 3.2.6** ***(Size–biased sampling).*** Consider the pdf $g(x) = C\cdot xf(x)$, $f \in \mathcal{T}$, corresponding to sampling from a GGC with probability proportional to size x. It is assumed that $\int xf(x)dx < \infty$ but if that integral is not finite, f(x) could be multiplied by $e^{-\theta x}$. Is g(x) also a GGC? The mgf of g equals

$$\int e^{sx}g(x)dx = C\cdot\varphi'(s) = C\cdot\frac{\varphi'(s)}{\varphi(s)}\cdot\varphi(s).$$

Returning to the specific distribution in Example 3.2.3, we have

$$C\cdot\varphi'(s)/\varphi(s) = \frac{\sqrt{1-\alpha}}{\sqrt{1-\alpha-s}}\cdot\frac{\sqrt{1+\alpha}}{\sqrt{1+\alpha-s}}.$$

Hence, in this case g equals the convolution of f with two Gamma densities and is therefore a GGC. This conclusion holds also for many other GGC's but not for all. However, g is always ID since

$$C\cdot\varphi'(s)/\varphi(s) = \int \frac{t}{t-s}\cdot\frac{C}{t}\,U(dt)$$

is the mgf of an MED; cf. Section 2.4.

The class of pdf's f(x) for which $C\cdot x^n e^{-\theta x}f(x)$ is a GGC for every $n \in \mathbb{N}_0$ will be identified in Chapter 5. □

**Example 3.2.7** ***(More size–biased sampling; a complicated example).*** Consider the pdf $f(x) = C\cdot(e^{-\lambda x} - e^{-x})$, $x > 0$, $0 < \lambda < 1$. It corresponds to the convolution of two Exponential distributions and is hence a GGC. Let $\tilde{f}(x) = C\cdot x^{-\gamma}f(x)$, $0 < \gamma < 2$, corresponding to size–biased sampling from f(x) with weight–function $x^{-\gamma}$. A surprising fact is that $\tilde{f}$ is also a GGC as we shall see. The mgf of $\tilde{f}$ can be shown to be given by

$$\tilde{\varphi}(s) \propto \begin{cases} (\lambda-s)^c - (1-s)^c, & c = \gamma-1,\ \gamma < 1 \\ -\log(\lambda-s) + \log(1-s), & \gamma = 1 \\ (1-s)^c - (\lambda-s)^c, & c = \gamma-1,\ \gamma > 1 \end{cases}$$

The constant of proportionality is irrelevant for what follows and is set to be 1. Obviously $\tilde{\varphi}(s)$ is analytic and zero–free in $\mathbb{C}\backslash[0,\infty)$. It is also easily verified that $\tilde{\varphi}'(s)/\tilde{\varphi}(s) \to 0$ uniformly as $|s| \to \infty$, $s \in \mathbb{C}\backslash[0,\infty)$. Consider first the case $\gamma > 1$. Then $\tilde{\varphi}(s)$ is also continuous up to and on the cut and the singularities of $\tilde{\varphi}'(s)/\tilde{\varphi}(s)$ at

$s = \lambda$ and $s = 1$ are of the harmless type mentioned in Remark 1 following Theorem 3.1.4. Further, for $t > 0$, we get by some calculation

$$\tilde{u}(t) = \frac{1}{\pi}\cdot \Im m[\tilde{\varphi}'(t)/\tilde{\varphi}(t)] = \begin{cases} \frac{1}{\pi}\cdot\frac{(1-\lambda)\cdot c\cdot \sin(c\pi)}{|\tilde{\varphi}(t)|^2}(t-\lambda)^{c-1}(1-t)^{c-1}, & \lambda < t < 1 \\ 0 & \text{elsewhere} \end{cases}$$

Here $\tilde{u}(t)$ is nonnegative and hence the distribution is a GGC by Theorem 3.1.4 and its Remark 1. Since

$$\tilde{\varphi}(s) = (-s)^c\cdot[(1+\frac{1}{-s})^c - (1+\frac{\lambda}{-s})^c] \sim c\cdot(1-\lambda)(-s)^{c-1}$$

as $|s| \to \infty$, $s \in \mathbb{C}\backslash[0,\infty)$, we have $\int \tilde{U}(dt) = \pi^{-1}\arg[\tilde{\varphi}(+\infty)] = 1-c = 2-\gamma$. An application of Formula (3.1.4) gives the same result.

For $\gamma < 1$, $\tilde{\varphi}'(s)/\tilde{\varphi}(s)$ has influential singularities at $s = \lambda$ and $s = 1$. We have for $s_0 = \lambda$ and $s_0 = 1$, $\tilde{\varphi}'(s)/\tilde{\varphi}(s) \sim -c/(s_0-s)$ as $s \to s_0$, $\Im m\, s > 0$. However, for $t \neq \lambda$ and $t \neq 1$, the expression for $\tilde{u}(t)$ above is still valid. Applying Cauchy's integral formula to $\tilde{\varphi}'(s)/\tilde{\varphi}(s)$, we may just integrate around the singularities along half–circles and get

$$\tilde{\varphi}'(s)/\tilde{\varphi}(s) = \int \frac{1}{t-s}\tilde{u}(t)dt + \frac{-c}{\lambda-s} + \frac{-c}{1-s}\,.$$

Again the distribution is a GGC as $-c > 0$. It has two Gamma distributions as convolution factors. Since now $\tilde{\varphi}(s) \sim -c\cdot(1-\lambda)(-s)^{c-1}$ as $|s| \to \infty$, we have, as before, $\int \tilde{U}(dt) = 1-c = 2-\gamma$. However, now $\int \tilde{u}(t)dt = 1-c+2c = \gamma$. The same conclusion can be verified to hold for $\gamma = 1$ also.

The very particular result in this example will be considerably generalized in Section 4.4 and Chapter 6. □

### 3.3 A Composition Theorem

In this section and in Section 3.4 the mgf will continue to be in focus instead of the pdf. To reach more central matters rapidly, the reader may here pass directly to Chapter 4 and return to this section when needed.

In Section 2.3 we saw that when $X(\tau)$ is a Lévy process with mgf $\exp\{\tau\cdot\zeta(s)\}$ and $T \geq 0$ is an independent ID random time with mgf $\varphi$, then $X(T)$ is ID with mgf $\varphi(\zeta(s))$. Let now $T \sim \mathcal{T}$ and let $X(\tau)$ be a $\mathcal{T}$–process, i.e., a Lévy process for which the increments have their distributions in $\mathcal{T}$. It is not true that in general $X(T) \sim \mathcal{T}$ but it holds under an additional condition on the $\mathcal{T}$-process, i.e., on $\zeta(s)$.

**Theorem 3.3.1** ***(Composition theorem).*** *Let $e^{\zeta(s)}$ be the mgf of a GGC ($\zeta(0) = 0$ and $\zeta$ is continuous). Then $\varphi(\zeta(s))$ is the mgf of a GGC whenever $\varphi$ is iff*

$$\Im m[\zeta'(s)\overline{\zeta(s)}] \leq 0 \quad \text{for } \Im m\, s > 0 . \tag{3.3.1}$$

**Proof.** Since

$$\zeta(s) = a's + \int \log(\tfrac{t}{t-s})\, V(dt), \quad a' \geq 0,\ V(dt) \geq 0,$$

and the function $s \to \log(\frac{t}{t-s})$ belongs to $\mathcal{P}(-\infty, 0)$, it follows that $\Im m[\zeta(s)] > 0$ for $\Im m\, s > 0$ (unless $\zeta(s) \equiv 0$). Hence, and since $\zeta(s) < 0$ for $s < 0$, we have, for $s \neq 0$, $\zeta(s) \in \mathbb{C}\backslash[0,\infty)$. Thus, for any mgf $\varphi$ of a GGC, the mgf $\psi(s) = \varphi(\zeta(s))$ is a well–defined analytic and zero–free function in $\mathbb{C}\backslash[0,\infty)$. Further,

$$\frac{\psi'(s)}{\psi(s)} = \frac{\varphi'(\zeta(s))}{\varphi(\zeta(s))}\cdot\zeta'(s) = a\zeta'(s) + \int \frac{\zeta'(s)}{t-\zeta(s)}\, U(dt) ,$$

where $a$ and U correspond to $\varphi$. Hence

$$\Im m[\frac{\psi'(s)}{\psi(s)}] = a\cdot\Im m[\zeta'(s)] + \int \frac{\Im m[\zeta'(s)\cdot(t-\overline{\zeta(s)})]}{|t-\zeta(s)|^2}\, U(dt) .$$

Since $e^{\zeta(s)}$ is the mgf of a GGC, certainly $\Im m[\zeta'(s)] \geq 0$ for $\Im m\, s > 0$. In view of Theorem 3.1.2, it is now immediately seen that (3.3.1) is a sufficient condition for $\psi$ to be the mgf of a GGC. The necessity of the condition follows from Theorem 3.1.3 by choosing $\varphi(s) = \frac{t}{t-s}$ and letting t tend to 0. □

It is not hard to verify that $\zeta(s) = \beta\cdot\log(\frac{t}{t-s})$, $\beta > 0$, $t > 0$, corresponding to a Gamma distribution, does not satisfy (3.3.1). Thus for a Gamma process $X(\cdot)$, it is not true that $X(T) \sim \mathcal{T}$ whenever $T \sim \mathcal{T}$. However, it holds true when $X(\cdot)$ is an $\mathcal{L}^{\infty}$–process, i.e., a superposition of Stable processes with different indices; cf. Section 2.3. We have:

**Theorem 3.3.2.** *Let $\zeta(s)$ be of the form $\zeta(s) = -\Sigma_{k=1}^{n} c_k(-s)^{\alpha_k}$, $c_k > 0$, $0 < \alpha_k \leq 1$, or a limit thereof. Then $\varphi(\zeta(s))$ is the mgf of a GGC whenever $\varphi$ is.*

**Proof.** Since Stable distributions are GGC's, $e^{\zeta(s)}$ is the mgf of a GGC. Only Condition (3.3.1) has to be verified. We use induction on n. The condition holds trivially for $n = 0$. Assume that (3.3.1) holds when the sum defining $\zeta(s)$ has n–1 terms. We shall prove that (3.3.1) holds also when the sum has $n$ terms. It is no restriction to assume that $\alpha_1 < \alpha_2 < \ldots < \alpha_n$. Since $\zeta(s) = -\Sigma_{k=1}^n c_k(-s)^{\alpha_k}$ is a Pick function, $\zeta(s) \neq 0$ for $s \neq 0$. Set $\psi(s) = -1/\zeta(s)$. Now

$$\psi(s) = \frac{1}{\Sigma_{k=1}^n c_k(-s)^{\alpha_k}} = \frac{1}{(-s)^{\alpha_1}} \cdot \frac{1}{\Sigma_{k=1}^n c_k(-s)^{\alpha_k - \alpha_1}} = \frac{1}{(-s)^{\alpha_1}} \cdot \chi(s) \quad \text{(say)}$$

For $n \geq 2$, $\chi(s)$ is proportional to the mgf of a GGC by the assumption and by Theorem 3.3.1; note that the first term of the second sum is $c_1$. For $n = 1$, $\chi(s) = 1/c_1$. Further,

$$\psi'(s)/\psi(s) = \alpha_1/(-s) + \chi'(s)/\chi(s) .$$

Hence and by Theorem 3.1.2, $\Im m[\psi'(s)/\psi(s)] \geq 0$ for $\Im m\, s > 0$. Since $\psi'/\psi = -\zeta'/\zeta$, it follows that $\Im m[\zeta'(s)\overline{\zeta(s)}] \leq 0$ for $\Im m\, s > 0$. □

The result in Example 3.2.4 is an obvious consequence of this theorem. Another less obvious consequence is now given.

**Example 3.3.1** ***(Exponential integral).*** The function $\varphi(s) = (-s)^{-1}\log(1-s)$ is the mgf of a GGC. We verify this assertion by using Theorem 3.3.2. We set

$$\zeta(s) = -\int_0^1 (-s)^{\alpha} d\alpha = \frac{1+s}{\log(-s)} .$$

By Theorem 3.3.2, $\psi(s) = \frac{t}{t-\zeta(s)}$ $(t > 0)$ corresponds to a GGC and hence so does

$$\frac{\psi(s-1)}{\psi(-1)} = \frac{t + 1}{t - \frac{s}{\log(1-s)}} .$$

Since this mgf tends to $\varphi(s)$ as $t \to 0$, the statement is proved. The corresponding pdf is the exponential integral $f(x) = E_1(x) = \int_1^\infty y^{-1}e^{-xy}dy$, as can be verified by calculation of $\int e^{sx}f(x)dx$ or by the inversion formula (3.1.9); cf., e.g., Springer (1979, pp. 177–178). It will be given more attention in Chapters 5 and 8. □

**Remark.** Let $\mathscr{Z}$ be the class of functions $\zeta$ such that $\varphi(\zeta(s))$ is the mgf of a GGC whenever $\varphi$ is. By its very definition, $\mathscr{Z}$ is closed under composition. For instance,

$$\zeta(s) = -\Sigma_{k=1}^n c_k \left(\Sigma_{j=1}^m d_j(-s)^{\beta_j}\right)^{\alpha_k},$$

with obvious restrictions on the parameters, belongs to $\mathscr{Z}$. On the other hand, it can be shown that for $m > 1$, this function is not in the subclass of functions in $\mathcal{Z}$ provided by Theorem 3.3.2. That subclass is closed under addition. There is no reason to expect $\mathcal{Z}$ also to be closed under addition. □

### 3.4. Two Real Characterizations of the Mgf of a GGC

In Section 3.1 the GGC's were characterized by the Pick function property of the derivative of the logarithm of the mgf. This *complex* characterization is very useful. In this section we give two *real* characterizations. They are not that useful but have some interest. The first one is actually a characterization for a sequence of nested classes of distributions.

**Theorem 3.4.1.** *Let $F$ be a probability distribution on $\mathbb{R}_+$ with mgf $\varphi$ and let $n \in \mathbb{N}_0$. Then $F$ is ID* **and** *$r(y)$ $(= y\ell(y)$, where $\ell$ is the density of the Lévy measure) satisfies*

$$(3.4.1) \qquad (-1)^k r^{(k)}(y) \geq 0\,, \quad y > 0, \quad k = 0, 1, \dots, n,$$

***iff**, on $(-\infty, 0)$,*

$$(3.4.2) \qquad \left(s^k \cdot \frac{\varphi'(s)}{\varphi(s)}\right)^{(k)} \text{ is absolutely monotone, } \quad k = 0, 1, \dots, n.$$

*The function $(-1)^n r^{(n)}(y)$ in (3.4.1) is permitted to be just a nonnegative measure, i.e. the derivative of an increasing function. Moreover, if (3.4.1) or (3.4.2) hold, then, for some $a \geq 0$,*

$$(3.4.3) \qquad s^k \cdot \frac{\varphi'(s)}{\varphi(s)} = as^k + \int \left(e^{sy} - \sum_{j=0}^{k-1} \frac{(sy)^j}{j!}\right) \cdot (-1)^k r^{(k)}(y)dy\,, \quad k = 0, 1, \dots, n.$$

In particular, in view of Theorem 3.1.1, $F \in \mathcal{T}$ iff $(s^k\varphi'(s)/\varphi(s))^{(k)}$ is absolutely monotone for all $k \in \mathbb{N}_0$. For $n = 0$ a well–known characterization of the ID distributions appears if the theorem is read correctly. For $n = 1$ a characterization of the self–decomposable distributions appears.

**Proof.** We first show that (3.4.1) implies (3.4.2) and (3.4.3). We proceed by induction on n. For $n = 0$ the theorem holds. *We assume that (3.4.1) holds for $k \leq n$ and that (3.4.2) and (3.4.3) hold for $k \leq n-1$.* It follows from (3.4.3) for $k = n-1$ that

$$\int_0^1 y^{n-1} (-1)^{n-1} r^{(n-1)}(y)dy < \infty \quad \text{and} \quad \int_1^\infty y^{n-2} (-1)^{n-1} r^{(n-1)}(y)dy < \infty\,,$$

and hence, since $(-1)^{n-1} r^{(n-1)}(y)$ is decreasing, that

$$y^n r^{(n-1)}(y) \to 0 \quad \text{as } y \to 0 \quad \text{and} \quad y^{n-1} r^{(n-1)}(y) \to 0 \quad \text{as } y \to \infty. \tag{3.4.4}$$

Multiplying (3.4.3) for $k = n-1$ by $s$ and then evaluating the integral by an integration by parts, we get in view of (3.4.4)

$$s^n \cdot \frac{\varphi'(s)}{\varphi(s)} = as^n + \int (e^{sy} - \sum_{j=0}^{n-1} \frac{(sy)^j}{j!}) \cdot (-1)^n r^{(n)}(y)\, dy,$$

where now $(-1)^n r^{(n)}(y)dy$ may be just a nonnegative measure on $(0,\infty)$. Hence (3.4.3) holds for $k = n$. Moreover,

$$(s^n \cdot \frac{\varphi'(s)}{\varphi(s)})^{(n)} = n!a + \int e^{sy} y^n (-1)^n r^{(n)}(y)\, dy,$$

i.e., also (3.4.2) holds for $k = n$. The proof of the "only if" part is complete.

Turning to the harder converse, we use induction again and *assume instead that (3.4.2) holds for $k \le n$ and that (3.4.1) and (3.4.3) hold for $k \le n-1$.* We want to prove that $(-1)^n r^{(n)}(y)dy$ is a nonnegative measure, i.e., that $(-1)^{n-1} r^{(n-1)}(y)$ is a decreasing function. Now we know that

$$\int_0^1 y^k (-1)^k r^{(k)}(y)dy < \infty \quad \text{and} \quad \int_1^\infty y^{k-1} (-1)^k r^{(k)}(y)dy < \infty \tag{3.4.5}$$

for $k = 0, 1, \ldots, n-1$. For simplicity, set $g(s) = s^{n-1}\varphi'(s)/\varphi(s)$. By Bernstein's theorem,

$$(sg(s))^{(n)} = c + \int e^{sy}\mu(dy), \quad c \ge 0, \quad \mu(dy) \ge 0.$$

It follows after an integration that

$$(sg(s))^{(n-1)} = cs + \int \frac{e^{sy}-1}{y}\,\mu(dy) \tag{3.4.6}$$

provided that $(sg(s))^{(n-1)} \to 0$ as $s \uparrow 0$. To verify that condition, we note that, by the assumptions,

$$(sg(s))^{(n-1)} = sg^{(n-1)}(s) + (n-1)g^{(n-2)}(s)$$

$$= s \cdot [b + \int e^{sy} y^{n-1} (-1)^{n-1} r^{(n-1)}(y)dy] + (n-1)[bs + \int (e^{sy}-1) y^{n-2} (-1)^{n-1} r^{(n-1)}(y)dy],$$

where $b = (n-1)!a$. By (3.4.5) and the fact that $sy \cdot e^{sy}$ is bounded (by $1/e$), it now easily follows that $(sg(s))^{(n-1)} \to 0$ as desired. Analogous repeated integrations of (3.4.6) with respect to s give

$$\text{(3.4.7)} \qquad sg(s) = s^n \cdot \frac{\varphi'(s)}{\varphi(s)} = \frac{cs^n}{n!} + \int \Big(e^{sy} - \sum_{j=0}^{n-1} \frac{(sy)^j}{j!}\Big) \frac{\mu(dy)}{y^n} .$$

Certainly the integrals $\int_{(0,1]} \mu(dy)$ and $\int_{(1,\infty)} \frac{\mu(dy)}{y}$ are finite, and it follows that

$$y^{n-1} \int_{(y,\infty)} \frac{\mu(dx)}{x^n} \to 0 \text{ as } y \to \infty \quad \text{and} \quad y^n \int_{(y,\infty)} \frac{\mu(dx)}{x^n} \to 0 \text{ as } y \to 0.$$

An integration by parts in (3.4.7) therefore gives

$$s^n \cdot \frac{\varphi'(s)}{\varphi(s)} = \frac{cs^n}{n!} + s \cdot \int \Big(e^{sy} - \sum_{j=0}^{n-2} \frac{(sy)^j}{j!}\Big) \cdot \Big(\int_{(y,\infty)} \frac{\mu(dx)}{x^n}\Big) \, dy .$$

In view of (3.4.3) for $k = n-1$, it now easily follows that $c = n!a$ and

$$\int_{(y,\infty)} \frac{\mu(dx)}{x^n} = (-1)^{n-1} r^{(n-1)}(y) , \qquad y > 0.$$

This ends the proof. □

A real characterization of functions of the form $\psi(s) = a + \int_{[0,\infty)} \frac{1}{t-s} U(dt)$, $a \geq 0$ and $U(dt) \geq 0$, can be found in Widder (1946, p. 366). In part, Theorem 3.4.1 is closely related to that characterization. By Theorem 3.4.1 it can again be verified that a Stable distribution is a GGC but few other distributions can be proved to be in $\mathscr{T}$ by the help of this theorem.

Another real characterization can be derived from a celebrated result of Loewner; see Donoghue (1974, Chapter XI). In an important special case, Loewner's result states that a nonnegative function $\psi(s)$, $s < 0$, is of the above form iff every matrix $\mathbf{M}$ formed by elements $m_{ij} = (\psi(s_i) - \psi(s_j))/(s_i - s_j)$ $(= \psi'(s_i)$ if $s_i = s_j)$ is nonnegative definite. A matrix is nonnegative definite if $\sum\sum m_{ij} x_i x_j \geq 0$, whatever the values of $x_1, x_2, \ldots, x_n$ are. The following theorem is an almost immediate consequence.

**Theorem 3.4.2.** *Let $F$ be a probability distribution on $\mathbb{R}_+$ with mgf $\varphi$. Then $F \in \mathscr{T}$ iff, for each $n \geq 2$ and each set of negative numbers $s_1, s_2, \ldots, s_n$, the $n \times n$ matrix $\mathbf{M}$ formed by the elements*

$$m_{ij} = \frac{\varphi'(s_i)\varphi(s_j) - \varphi'(s_j)\varphi(s_i)}{s_i - s_j}$$

*is nonnegative definite.*

Although some results can be obtained by this characterization it is not sufficiently powerful for our purposes. In Chapter 6 a more powerful real characterization is given.

### 3.5 Widened GGC's

We may give up the assumption that $\int F(dx) = 1$ and only require the measure $F(dx)$ to be a vague limit on $(0,\infty)$ of a sequence of convolutions of functions of the Gamma type or GGC's multiplied by positive constants. A nonnegative measure with mgf of the form

$$\varphi(s) = C\cdot\frac{1}{(-s)^{U(\{0\})}}\cdot\exp\{as + \int_{(0,\infty)}\log(\frac{t+1}{t-s})U(dt)\} ,$$

where $C > 0$, $a \geq 0$, and the nonnegative measure $U$ on $[0,\infty)$ now only satisfies $\int_{[0,\infty)}\frac{1}{1+t}U(dt) < \infty$, is said to belong to the widened class of GGC's, $\mathcal{T}_W$. Obviously $\varphi'(s)/\varphi(s) = a + \int_{[0,\infty)}\frac{1}{t-s}U(dt)$. Moreover, $F(dx) \in \mathcal{T}_W$ iff, for all $\epsilon > 0$, $e^{-\epsilon x}F(dx)$ can be normalized to be a GGC. In particular, Lebesgue measure with mgf $(-s)^{-1}$ is in $\mathcal{T}_W$ and so is the convolution of Lebesgue measure and a GGC with pdf f. This convolution has the density $F(x) = \int_0^x f(y)dy$ and mgf $(-s)^{-1}\varphi(s)$, where $\varphi$ corresponds to f. Thus, the cdf of a GGC is, as a density, in $\mathcal{T}_W$. More generally, if $f \in \mathcal{T}$, then the fractional integral (Euler transform of first kind)

$$\frac{1}{\Gamma(\gamma)}\cdot\int_0^x (x-y)^{\gamma-1}f(y)dy, \quad \gamma > 0,$$

with mgf $(-s)^{-\gamma}\varphi(s)$, is a density in $\mathcal{T}_W$. There is a simple converse as well.

In a very much widened GGC class there are also densities (measures) for which the mgf does not exist as a function. Examples are provided by $f(x) = x^{-3/2}$ and $f(x) = e^{-x}/x$ $(x > 0)$.

### 3.6 Bivariate GGC's

There are two ways of defining a bivariate GGC (BVGGC).

i) The distribution of a random vector $(X_1,X_2)$, where $X_j \geq 0$, $j = 1, 2$, is said to be a BVGGC *in the weak sense* if $c_1X_1 + c_2X_2 \sim \mathcal{T}$ whenever $c_1, c_2 \geq 0$.

ii) A distribution is said to be a BVGGC *in the strong sense* if it is the limit distribution for a sequence of random vectors of the type $(\Sigma_{j=1}^{n} c_{1j}Y_j , \Sigma_{j=1}^{n} c_{2j}Y_j)$, where the $Y_j$'s are independent Gamma variables (with unit scale) and the coefficients are nonnegative. The bivariate mgf $\varphi(s_1,s_2)$ is thus a pointwise limit of products with factors of the form $(1-c_1s_1-c_2s_2)^{v}$.

A canonical representation for a BVGGC in the strong sense can be shown to be

$$\varphi(s_1,s_2) = \exp\{a_1s_1+a_2s_2+\iint_{\mathbb{R}_+^2\setminus(0,0)} \log(\frac{1}{1-c_1s_1-c_2s_2})V(dc_1,dc_2)\},$$

where $a_1, a_2 \geq 0$ and $V(dc)$ is nonnegative. It is easy to see that the imaginary parts of $\frac{\partial\varphi}{\partial s_1}/\varphi$ and $\frac{\partial\varphi}{\partial s_2}/\varphi$ are nonnegative in the region $\{(s_1,s_2); \Im m\, s_1 > 0, \Im m\, s_2 > 0\}$, but unfortunately this nonnegativity does not seem to be a characteristic property.

The mgf $\varphi(s_1,s_2) = (1-s_1+\sqrt{-s_2}\,)^{-\gamma}$, $\gamma > 0$, corresponds to the random vector $(Y\cdot X_1, Y^2\cdot X_2)$, where $Y \sim \text{Gamma}(\gamma,1)$, $X_1 \equiv 1$, and $X_2$ is Stable of index 1/2. It follows easily from Theorem 3.3.2 that $\varphi$ is the mgf of a BVGGC in the weak sense. However, $\varphi$ is not the mgf of a BVGGC in the strong sense since $\Im m[\frac{\partial\varphi}{\partial s_2}/\varphi]$ is not nonnegative in the region above. On the other hand, since $\varphi$ is a product of any number of identical mgf's, it is ID in a strong sense. Distributions which are ID in the weak sense but not in the strong sense exist as well; see, e.g., Stoyanov (1987, pp. 72–74).

### 3.7 Complementary Results and Notes

I (***Behaviour on the cut***). The mgf $\varphi$ of a GGC may be defined on the upper (or lower) side of the cut but may behave wildly there. Below some facts are given.

a) If $U(t)$ is continuous for $t > 0$ and locally satisfies a Lipschitz condition of the type $|U(t')-U(t)| \leq C\cdot|t'-t|^{\epsilon}$ for some $\epsilon > 0$, then $\varphi$ is continuous up to and on the cut outside 0. This can be verified from the relation: $\int \log(\frac{t}{t-s})U(dt) = \int (\frac{1}{t-s} - \frac{1}{t})U(t)dt$.

b) Even if $U$ has a density $u$, the mgf $\varphi$ can be discontinuous at 0. That may happen if $u$ oscillates wildly in a neighbourhood of 0.

c) The mgf $\varphi$ has no zeros on the cut. In fact, $|\varphi(s)|$, $s \in \mathbb{C}\setminus[0,\infty)$, can never tend to zero as $s$ tends to a real value.

II (***Inversion***). Let $f$ and $\varphi$ be the pdf and the mgf, respectively, of a GGC with $a = 0$ and $\int U(dt) = \beta < \infty$. According to (3.1.9), we have $f(x) = \int_0^\infty e^{-xt}g(t)dt$, where

$$g(t) = \frac{1}{\pi}\cdot\Im m[\varphi(t)] = \frac{1}{\pi}\cdot\exp\{\int_0^\infty \log|\frac{y}{y-t}| U(dy)\}\cdot\sin(\pi U(t)), \quad t > 0.$$

Obviously $g(t) \geq 0$ for $\beta \leq 1$. Thus, for $\beta \leq 1$ the GGC is an MED as was also found at the end of Example 3.2.3. For $\beta > 1$, we have (as O. Thorin has communicated)

$$f(x) = x^{\beta-1}\cdot\int_0^\infty e^{-xt} \frac{1}{\Gamma(\beta-1)} \int_0^t (t-v)^{\beta-2} g(v)dv\, dt .$$

In fact, the mgf's of two sides can be shown to be equal. The inner v–integral is nonnegative but that is hard to verify. It will be a consequence of Theorem 4.1.1.

III (***Curious composition***). Let $\varphi$ be the mgf of a GGC with $a = 0$ and $\int U(dt) = \beta < \infty$. Then $\psi(s) = (-s)^{-\beta}\varphi(1/s)$ is also the mgf of a GGC with total U–mass $\beta$. In fact, as easily can be verified, the new U–measure is induced by U by the transformation $t \to 1/t$. It follows that $(-s)^{-\beta\alpha}\varphi(-1/(-s)^{\alpha})$, $0 < \alpha \leq 1$, is also the mgf of a GGC.

IV (***Complete GGC***). Berg & Forst (1983) have modified Urbanik's result, cf. Section 2.3, as follows: Let here $\mathcal{T}^0 = \mathcal{T}$. Defining then iteratively $\mathcal{T}^k$ $(k = 1, 2, ...)$ as the class of distributions in $\mathcal{T}^{k-1}$ for which $\epsilon_c \sim \mathcal{T}^{k-1}$, cf. Section 2.3, and setting $\mathcal{T}^\infty = \cap\, \mathcal{T}^k$, they showed that $\mathcal{T}^\infty = \mathcal{L}^\infty$. The relation to Theorem 3.3.2 is not clear. Moreover, it is not known whether that theorem is valid with GGC's replaced by self–decomposable distributions.

**Notes.** The GGC's were introduced by Thorin (1977a,b) as a tool for proving the ID of the Lognormal distribution. Thorin's field was insurance mathematics. The Lognormal distribution has been used for modelling periodic total claims on insurance companies. If this is a good model, the distribution should be at least ID. Steutel (1973) mentioned the ID of the Lognormal distribution as an open problem. Theorems 3.1.1, 3.1.4 (Inversion Theorem), and 3.1.5 (Closure Theorem) are due to Thorin.

Theorems 3.1.2 and 3.1.3 were given in B1981. Example 3.2.1 (Stable distribution) is due to Thorin (1977a). The result in Example 3.2.3 (First passage time) improves on Feller's (1966) corresponding ID result. See also Steutel (1970, p. 68). (Feller was apparently very fond of his result; he returns to it several times in his (1971) book.) The remaining examples are essentially new.

Theorems 3.3.1 (Composition Theorem) and 3.3.2 date in part back to B1979a,c and B1988a but at least the proofs here are new. Theorem 3.4.1 (Real characterization) is new. The special case $n = 1$, corresponding to self–decomposability, has been treated before in greater generality; see Wolfe (1982), Jurek & Vervaat (1983), and Forst (1984).

The widened GGC's and the bivariate GGC's have not been treated before.

# 4. DENSITIES OF GENERALIZED GAMMA CONVOLUTIONS

In Chapter 3 the distributions in $\mathcal{T}$ were studied through the mgf. In this chapter we look at densities of GGC's and their properties. In particular, classes of densities in $\mathcal{T}$ are in focus. Some rather surprising results appear.

In Section 4.1 it is shown that a GGC is a mixture of Gamma distributions. In Section 4.2 a large class of mixtures of Exponential distributions (MED) is found that is a subclass of $\mathcal{T}$. The class of *hyperbolically completely monotone* (HCM) densities is introduced in Section 4.3. Densities in this class are shown to be in $\mathcal{T}$. In examples some well–known pdf's are verified to be HCM and thus in $\mathcal{T}$. Section 4.4 treats size–biased sampling from GGC's.

## 4.1 GGC's as Mixtures of Gamma Distributions

We saw at the end of Example 3.2.3 that every GGC with $a = 0$ and $\beta = \int U(dt) \leq 1$ is an MED. There is a generalization stating that a GGC with $a = 0$ and $0 < \beta < \infty$ is a mixture of Gamma($\beta,\cdot$)–distributions.

**Theorem 4.1.1** *(Gamma mixture representation). The pdf $f$ of a GGC with $a = 0$ and $\int U(dt) = \beta$, $0 < \beta < \infty$, admits the representation $f(x) = x^{\beta-1}h(x)$, where $h$ is completely monotone (CM). Equivalently, if $X \sim \mathcal{T}$ (with $a$ and $\beta$ as above), then $X \overset{d}{=} Y\cdot Z$, where $Y \sim$ Gamma($\beta$,1) and $Z > 0$ are independent rv's.*

**Proof.** The equivalence of the two statements is rather immediate and is left to the reader to verify; $h(x) = E[Z^{-\beta}e^{-x/Z}]/\Gamma(\beta)$. To prove the second statement, assume first that the distribution of $X$ is a finite convolution of $n$ Gamma distributions with shape parameters $\beta_1, ..., \beta_n$, the sum of which is $\beta$. Then $X \overset{d}{=} \Sigma\, c_jX_j$, where $c_j > 0$ and $X_j \sim$ Gamma($\beta_j$,1), $j = 1, ..., n$, are independent rv's. Equivalently, $X \overset{d}{=} Y_n\cdot Z_n$, where $Y_n = \Sigma\, X_j$ and $Z_n = \Sigma\, c_jV_j$ with $V_j = X_j/Y_n$. Now, as is well known, $Y_n \sim$ Gamma($\beta$,1) and $Y_n$ is independent of the vector $(V_1, ..., V_n)$. Hence $Y_n$ and $Z_n$ are independent and the theorem holds in the special case of finite convolutions. Turning to the general case, we first note that by the converse part of the Closure Theorem and some reflection, the distribution of $X$ can be arbitrarily closely approximated by finite convolutions of Gamma distributions with shape parameters $\beta_{jn}$, $j = 1, ..., n$, the sum of which is $\beta$. The U–measure is just appropriately discretized. Hence and as above, $Y_n\cdot Z_n \to X$ in distribution as $n \to \infty$. The sequence of distributions of $Z_n$, $n = 2, 3, ...$ , can now be thinned so that it converges weakly to some probability distribution on $[0,\infty]$ at least. Let Z be an (extended) rv with that distribution and let $Y \sim \text{Gamma}(\beta,1)$ be independent of Z. Since, after the thinning, $(Y_n,Z_n) \to (Y,Z)$ in distribution, it follows that $X \overset{d}{=} Y\cdot Z$. Since $0 < X,Y < \infty$, we have $0 < Z < \infty$. □

There is no direct converse of Theorem 4.1.1. However, we shall see in Chapter 6 that for independent rv's $Y \sim$ Gamma and $Z \sim \mathcal{T}$, necessarily $Y\cdot Z \sim \mathcal{T}$.

The CM function h in Theorem 4.1.1 may be infinite at $x = 0$.

**Theorem 4.1.2.** *Let $f$ be the pdf of a GGC with $a = 0$ and $\int U(dt) = \beta$, $0 < \beta < \infty$. Then $h(x) = f(x)/x^{\beta-1}$ is finite at $x = 0$ iff $\int_{(1,\infty)}(\log t)U(dt) < \infty$. In this case*

$$h(0+) = \frac{1}{\Gamma(\beta)}\cdot exp\{\int_{(0,\infty)}(\log t)U(dt)\}.$$

**Proof.** Let $\phi(s) = \varphi(-s)$ be the Laplace transform (LT) of f. Now, as $s \to \infty$,

$$s^\beta\phi(s) = s^\beta \int e^{-sx}x^{\beta-1}h(x)dx = \int e^{-y}y^{\beta-1}h(\tfrac{y}{s})dy \sim h(0+)\Gamma(\beta).$$

On the other hand,

$$s^\beta\phi(s) = \exp\{\int \Big(\log(\tfrac{t}{t+s}) + \log s\Big)\, U(dt)\}$$

$$= \exp\{\int \Big(\log t + \log(\tfrac{s}{t+s})\Big)U(dt)\} \sim \exp\{\int (\log t)U(dt)\}, \quad \text{as } s \to \infty.$$

This proves the assertion. □

Since h is CM, the pdf f is differentiable infinitely many times on $(0,\infty)$. This is true also when $\beta = \infty$. More generally:

**Theorem 4.1.3.** *Let $f$ be the pdf of a GGC with $a = 0$ and $\int U(dt) = \beta$, $1 < \beta \leq \infty$. Let $k \geq 0$ be an integer such that $k < \beta-1$. Then $f$ is continuously differentiable any number of times on $(0,\infty)$ and at $0$ at least $k$ times with $f^{(j)}(0) = 0$ for $j \leq k$.*

**Proof.** Let $\varphi(is)$, $s \in \mathbb{R}$, be the characteristic function of f. Then (cf. the proof of Theorem 4.1.2)

$$(-is)^\beta\varphi(is) \to \exp\{\int (\log t)U(dt)\} \quad \text{as } s \to \pm\infty.$$

Hence, if U has bounded support, then $|\varphi(is)| = O(|s|^{-\beta})$ as $s \to \pm\infty$. If U has unbounded support, then

$$(4.1.1) \qquad |\varphi(is)| = O(|s|^{-\beta_1}), \quad s \to \pm\infty,$$

for any $\beta_1 < \beta$. In fact, U can be split into two measures, $U_1$ and $U_2$, such that $U_1$ has total mass $\geq \beta_1$ and that mass in a finite interval $(0,A]$ and $U_2$ has its mass in $(A,\infty)$. Then $\varphi(is) = \varphi_1(is)\varphi_2(is)$ and since (4.1.1) holds for $\varphi_1$ and a characteristic function is bounded in modulus by 1, it follows that (4.1.1) holds. Now, for $\beta > 1$, $\varphi(is)$ is absolutely integrable and hence

$$f(x) = \frac{1}{2\pi}\int_{-\infty}^{\infty} e^{-isx}\varphi(is)ds, \quad x \in \mathbb{R},$$

see, e.g., Feller (1971, p. 509). Moreover, in view of (4.1.1), the integral can be differentiated k times $(k < \beta-1)$ and each derivative is continuous on $\mathbb{R}$. Certainly $f^{(j)}(0) = 0$ for $j \leq k$ since $f(x) = 0$ for $x < 0$. □

A simple consequence of Theorem 4.1.3 is that if $f(0+) > 0$, then $\beta \leq 1$ and hence f is decreasing. If $f(0+) = 0$, then f is first increasing and then decreasing by the unimodality of a GGC. As mentioned in Chapter 3, the behaviour of f at $x = 0$ is related to the total mass of U in the following way:

**Theorem 4.1.4.** *For a non-degenerate GGC with $a = 0$ we have*

$$(4.1.2) \qquad \int U(dt) = \sup\{\alpha\, ;\, \lim_{x\downarrow 0} f(x)/x^{\alpha-1} = 0\}.$$

**Proof.** If $\beta = \int U(dt) = \infty$, then, for any $n \in \mathbb{N}_0$, by a McLaurin expansion of f at $x = 0$ and Theorem 4.1.3, $f(x)/x^n \to 0$ as $x \downarrow 0$. Hence (4.1.2) holds in this case. Now let $\beta < \infty$. Consider again the LT $\phi(s)$. Now, for $y > 0$, as $s \to \infty$,

$$\frac{\phi(sy)}{\phi(s)} = \exp\{\int \log(\tfrac{t+s}{t+sy})\, U(dt)\} \to \exp\{\int \log(y^{-1})\, U(dt)\} = y^{-\beta}.$$

Hence $\phi$ is *regularly varying* with exponent $-\beta$ at $s = \infty$; see, e.g., Feller (1971, pp. 442–448) or Bingham, Goldie & Teugels (1987, Chapter 1). Since f is unimodal and therefore ultimately monotone in a positive neighbourhood of $x = 0$, it follows that $h(x) = f(x)/x^{\beta-1}$ is *slowly varying* at $x = 0$, i.e., $h(xy)/h(x) \to 1$ as $x \downarrow 0$ for any $y > 0$. Then, as is known, for any $\epsilon > 0$, $x^{\epsilon}h(x) \to 0$ as $x \downarrow 0$. Hence, for any $\alpha < \beta$, $f(x)/x^{\alpha-1} \to 0$ as $x \downarrow 0$. Further, for $\alpha > \beta$, $f(x)/x^{\alpha-1} = x^{\beta-\alpha}h(x)$ which obviously tends to $\infty$ as $x \downarrow 0$ since, by the complete monotonicity of h, $h(0+) > 0$. This completes the proof. □

**Remark.** As a CM function $h(x)$ can be extended to an analytic function in $\Re e\, x > 0$, the same holds for the pdf $f(x) = x^{\beta-1}h(x)$ of a GGC with $a = 0$ and $0 < \beta < \infty$. If the measure U has its mass in an interval $(0, t_0]$, $t_0 < \infty$, more can be said. Returning to Theorem 4.1.1 and its proof, we see that now the rv Z takes its values in $[t_0^{-1}, \infty)$. It follows that $h(x) = E[Z^{-\beta}e^{-x/Z}]/\Gamma(\beta)$ is analytic in the whole complex plane $\mathbb{C}$. Hence $f(x)$ can be extended to an analytic function in $\mathbb{C}\backslash(-\infty,0]$. It is unknown whether this is true for every non–degenerate GGC with $a = 0$. □

Let $f_n$, $n = 1, 2, \ldots$ , be a sequence of densities of GGC's. If $f_n \to f$ pointwise, where also f is a pdf, then, by Scheffé's theorem, the convergence is weak as well. There is also a converse.

**Theorem 4.1.5.** *Let $f_n \in \mathcal{T}$, $n = 1, 2, \ldots$ , converge to $f \in \mathcal{T}$ weakly. Let $a$ and $U$ be the left–extremity and the Thorin measure, respectively, corresponding to $f$.*

*a) If $\int U(dt) > 1$, then $f_n(x) \to f(x)$ for all $x \in \mathbb{R}$.*

*b) If $\int U(dt) \leq 1$, then $f_n(x) \to f(x)$ for all $x \neq a$.*

**Proof. (a)** Since $U_n \to U$ vaguely on $(0,\infty)$, there exist a compact set $K \subset (0,\infty)$, a number $u_0 > 1$, and an $n_0 \in \mathbb{N}$ such that $\int_K U_n(dt) \geq u_0$ for all $n \geq n_0$. It follows easily that for some $C > 0$ and $t_0 > 0$

$$|\varphi_n(is)| \leq C\cdot(t_0+|s|)^{-u_0}, \quad s \in \mathbb{R},\ n \geq n_0.$$

Hence, by dominated convergence,

$$f_n(x) = \frac{1}{2\pi}\int_{-\infty}^{\infty} e^{-isx}\varphi_n(is)ds \to \frac{1}{2\pi}\int_{-\infty}^{\infty} e^{-isx}\varphi(is)ds = f(x).$$

**(b)** It is well known that the sequence of unimodality points of the pdf's $f_n$ tend to the unimodality point of f; see, e.g., Dharmadhikari & Joag–dev (1988, p. 3). The unimodality point of f is $a$ since $\int U(dt) \leq 1$. Hence, for an arbitrary $\epsilon > 0$, $f_n$ is increasing for $x < a-\epsilon$ and decreasing for $x > a+\epsilon$ if n is just sufficiently large. Let $x > a+\epsilon$. Then, for a small $\delta > 0$,

$$f_n(x) \geq \frac{F_n(x+\delta)-F_n(x)}{\delta} \to \frac{F(x+\delta)-F(x)}{\delta} \geq f(x+\delta).$$

Hence $\liminf f_n(x) \geq f(x+\delta)$. Similarly, $\limsup f_n(x) \leq f(x-\delta)$ if $x-\delta > a+\epsilon$. As $\delta$ and $\epsilon$ are arbitrary, $f_n(x) \to f(x)$ for $x > a$. The case $x < a$ is treated analogously. □

## 4.2 A Class of Mixtures of Exponential Distributions in $\mathcal{T}$

We saw in Section 4.1 that the pdf of a GGC with left–extremity 0 and $\int U(dt) = \beta < \infty$ necessarily is of the form $f(x) = x^{\beta-1}h(x)$, where h is CM. Certainly this form of a pdf f does not guarantee it to be in $\mathcal{T}$. The rest of this chapter is essentially devoted to the problem of finding sufficient conditions. In this section we look at the case $\beta = 1$ (or $\beta \leq 1$) corresponding to MED's. The following theorem has some nice consequences.

**Theorem 4.2.1.** *A pdf on $(0,\infty)$ of the form $f(x) = \int_0^\infty te^{-xt}m(t)dt$, with $m(t)$, or less restrictively $g(t) = tm(t)$, logconcave, is a GGC.*

**Proof.** Certainly $g(t)$ is logconcave if $m(t)$ is. Hence it suffices to show that logconcavity of g implies that $f \in \mathcal{T}$. Now, the mgf $\varphi$ is given by $\varphi(s) = \int \frac{1}{t-s} g(t)dt$ and obviously it is analytic in $\mathbb{C}\backslash[0,\infty)$. Recalling Theorem 3.1.3, we only have to verify that $\Im m[\varphi'(s)\overline{\varphi(s)}] \geq 0$ for $\Im m\, s > 0$. By a differentiation under the integral sign and an integration by parts,

$$\varphi'(s) = \int \frac{1}{(t-s)^2} g(t)dt = \int \frac{1}{t-s} g'(t)dt .$$

We have here used the logconcavity to ensure that g is the integral of its derivative g', which is well defined a.e., and to ensure that $g(t) \to 0$ as $t \to 0$ and $g(t)/t \to 0$ as $t \to \infty$; note that $\int t^{-1}g(t)dt = 1$. Now, for $\Im m\, s > 0$,

$$\Im m[\varphi'(s)\overline{\varphi(s)}] = \Im m[\int\int \frac{1}{t-s}\cdot\frac{1}{u-\bar{s}} g'(t)g(u)\, dtdu\,]$$

$$= \Im m\, s \cdot \int\int \frac{u-t}{|t-s|^2|u-s|^2} g'(t)g(u)\, dtdu$$

$$= \Im m\, s \cdot \int\int_{t<u} \frac{u-t}{|t-s|^2|u-s|^2} [g'(t)g(u)-g'(u)g(t)]\, dtdu \geq 0$$

since $g'(t)/g(t) \geq g'(u)/g(u)$ for $t < u$ due to the logconcavity. □

The U–measure is not easy to calculate explicitly for the GGC's provided by Theorem 4.2.1. If $\varphi(s)$ is smooth up to and on the cut $[0,\infty)$ in $\mathbb{C}$, the density of U at $s > 0$ equals

$$U'(s) = \frac{1}{\pi}\cdot\frac{\Im m[\varphi'(s)\overline{\varphi(s)}]}{|\varphi(s)|^2}, \tag{4.2.1}$$

where, on the right hand side, s is assumed to be on the upper side of the cut. Assuming that g and g' are at least continuous, we have, for $s > 0$,

$$\Re e[\varphi(s)] = \lim_{\epsilon\to 0} \int_{|t-s|>\epsilon} \frac{1}{t-s}\, g(t)dt \quad \text{and} \quad \Im m[\varphi(s)] = \pi\cdot g(s).$$

Replacing here g by g', we get expressions for $\Re e[\varphi'(s)]$ and $\Im m[\varphi'(s)]$. It follows that

$$\begin{aligned} \Im m[\varphi'(s)\overline{\varphi(s)}] &= \Im m[\varphi'(s)]\cdot\Re e[\varphi(s)] - \Re e[\varphi'(s)]\cdot\Im m[\varphi(s)] \\ &= \pi \int \frac{g'(s)g(t) - g'(t)g(s)}{t-s}\, dt\,, \end{aligned} \tag{4.2.2}$$

which is nonnegative at least when g is logconcave. In fact, the integral here can sometimes be evaluated. However, for calculation of $U'(s)$ also an explicit expression for $|\varphi(s)|^2 = (\Re e[\varphi(s)])^2 + (\Im m[\varphi(s)])^2$ is required and that is often hard to find.

**Example 4.2.1** ***(Pareto distribution).*** Let $m(t)$ be a Gamma$(\gamma,\lambda)$–density. Then $g(t) = \frac{1}{\Gamma(\gamma)}\lambda^\gamma t^\gamma e^{-\lambda t}$ which is logconcave since $g'(t)/g(t) = \gamma\cdot t^{-1} - \lambda$ is decreasing. The corresponding pdf $f(x)$ equals

$$f(x) = \int e^{-xt}g(t)\, dt = \gamma\lambda^\gamma\cdot(x+\lambda)^{-\gamma-1}, \quad x > 0.$$

Hence, by Theorem 4.2.1, the Pareto distribution is a GGC. Now, for $s > 0$, by (4.2.2),

$$\begin{aligned} \frac{1}{\pi}\cdot\Im m[\varphi'(s)\overline{\varphi(s)}] &= \int \frac{g'(s)/g(s) - g'(t)/g(t)}{t - s}\, g(s)g(t)\, dt \\ &= \int \frac{\gamma}{st}\, g(s)g(t)\, dt = \gamma\cdot\frac{g(s)}{s} = \gamma\cdot m(s). \end{aligned}$$

However, there is no simple expression for the density $U'(s)$. □

**Example 4.2.2** ***(More mixtures).*** Let $(\lambda_1, \lambda_2)$ be an interval contained in $(0, \infty)$ and let $g(t) = C > 0$ on that interval and zero elsewhere. Certainly $g(t)$ is logconcave and $m(t) = g(t)/t$ is a pdf if $C = 1/\log(\lambda_2/\lambda_1)$. It follows that

$$f(x) = \int e^{-xt}g(t)dt = C\cdot x^{-1}\cdot(e^{-\lambda_1 x} - e^{-\lambda_2 x})$$

is a GGC. We also get $\varphi(s) = C\cdot\log(\frac{\lambda_2-s}{\lambda_1-s})$ and, after some calculation,

$$\frac{1}{\pi}\cdot\Im m[\varphi'(s)\overline{\varphi(s)}] \;=\; C^2\cdot\frac{\lambda_2-\lambda_1}{(s-\lambda_1)(\lambda_2-s)}\,, \quad \lambda_1 < s < \lambda_2,$$

(and zero elsewhere). In this case there is an explicit expression for the density of U, but still the denominator in (4.2.1) is more complicated than the numerator. This result was essentially obtained already in Example 3.2.7. □

As mentioned in Section 2.1, logconcavity is preserved under convolution. This implies that the class of densities in $\mathcal{T}$ provided by Theorem 4.2.1 with logconcave g, like the class of CM densities, is closed with respect to multiplication (with proper normalization). In particular, it follows that a pdf of the form

$$f(x) = C\cdot \prod_{j=1}^{n} (1+c_j x)^{-\gamma_j}\,, \quad x > 0 \quad (c_j > 0,\ \gamma_j \geq 1)$$

is in $\mathcal{T}$. As will be seen in Chapter 5, this result can be generalized considerably.

Another consequence of Theorem 4.2.1 is worth mentioning. Let $f \in \mathcal{T}$ have left–extremity 0 and a logconcave density of its U–measure. Then, if $\int xf(x)dx < \infty$, the pdf $g(x)= C\cdot xf(x)$ also belongs to $\mathcal{T}$. In fact, since $\varphi'(s)/\varphi(s) = \int \frac{U'(t)}{t-s}\,dt$ is proportional to the mgf of a GGC, the result follows as in Example 3.2.6.

## 4.3 A Fundamental Class of Densities in $\mathcal{T}$

This section contains a result which is of very great importance. It has many remarkable consequences to be fully revealed in Chapters 5 – 8.

We start with a curious definition:

**Definition.** *A nonnegative function f on $(0, \infty)$ is called hyperbolically completely monotone (HCM) if, for each $u > 0$, $f(uv)f(u/v)$ is CM as a function of* $w = v + v^{-1}$.

It is easy to see that $f(uv)f(u/v)$ is always a function of w. By Bernstein's theorem, f is HCM iff there exists a family $\{K(d\lambda;u),\ u > 0\}$ of nonnegative measures on $[0, \infty)$ such that

$$\text{(4.3.1)} \qquad f(uv)f(u/v) \;=\; \int \exp\{-\lambda u\cdot(v+v^{-1})\}\; K(d\lambda;u).$$

(The extra factor u in the exponent simplifies subsequent formulae.)

Theorem 4.3.1 below essentially states that a pdf f which is HCM must be a GGC. Before the theorem is formulated, an heuristic proof is given in order to provide the reader with a good feeling for what the HCM–condition does and how it enters.

***Heuristic Proof.*** Let $\varphi$ be the mgf of f. Now

$$J(s) = \varphi'(s)\overline{\varphi(s)} = \int\int x \cdot e^{sx+\bar{s}y} f(x)f(y)dxdy .$$

Making the *hyperbolic substitution* $x = uv$, $y = u/v$, with Jacobian $-2u/v$, and using the representation (4.3.1), we find

$$\begin{aligned} J(s) &= \int 2u^2\Big(\int \exp\{suv+\bar{s}u/v\}\Big(\int \exp\{-\lambda uv-\lambda u/v\}K(d\lambda;u)\Big)dv\Big)du \\ &= \int 2u^2\Big(\int \Big(\int \exp\{-u(\lambda-s)v-u(\lambda-\bar{s})/v\}dv\Big)\ K(d\lambda;u)\Big)du . \end{aligned}$$

The *formal* substitution $v = u(\lambda-\bar{s})\rho$ (with $\rho > 0$ !) shows that the interior v–integral equals

$$\int u(\lambda-\bar{s})\cdot\exp\{-u^2|\lambda-s|^2\rho - \tfrac{1}{\rho}\}\ d\rho.$$

The imaginary part of the integrand is positive for $\Im m\ s > 0$ and hence so is $\Im m[J(s)]$. By Theorem 3.1.3, $f \in \mathcal{T}$. □

The formal substitution is dubious. Moreover, the heuristic proof only shows that $\Im m[J(s)] > 0$ for $\Re e\ s < 0$, $\Im m\ s > 0$. It seems as though natural additional conditions must be imposed on f to get a proof that works. Such conditions are now given.

According to Theorem 4.1.1 and Bernstein's theorem, the pdf of a GGC with left–extremity 0 and total U–measure $\beta$ $(0 < \beta < \infty)$ can be represented as

$$f(x) = x^{\beta-1} \int e^{-xt}G(dt), \quad G(dt) \geq 0.$$

Since f is integrable, $G(\{0\}) = 0$. It follows that $f(x) \to 0$ as $x \to \infty$ and $xf(x) \to 0$ as $x \to 0$. Moreover,

a) f can be extended to an analytic function $f(z)$ in $\Re e\ z > 0$.

b) In every sector $S(\epsilon) = \{z\ ;\ -\pi/2 + \epsilon \leq \arg[z] \leq \pi/2 - \epsilon\}$ $(\epsilon > 0)$, $f(z) \to 0$ uniformly as $|z| \to \infty$

c) In every sector $S(\epsilon)$, $zf(z) \to 0$ uniformly as $|z| \to 0$ $(z \neq 0)$.

These regularity conditions might hold also when $\beta = \infty$. The main theorem of this section and chapter, anticipated above, is now formulated.

**Theorem 4.3.1** ***(Main HCM–theorem).*** *Let $f$ be a pdf on $(0,\infty)$ such that (a), (b), and (c) hold. If, moreover, $f$ is HCM, then $f \in \mathcal{T}$ and has left–extremity 0. The total U–measure is given by (4.1.2) and may be $\infty$.*

To prove Theorem 4.3.1, a slight sharpening of (4.3.1) is needed.

**Lemma.** *If $f$ is HCM and $f(z)$ is analytic for $\Re e\, z > 0$, then (4.3.1) is valid for all complex $v$ such that $\Re e\, v > 0$.*

**Proof.** Consider $g(v) = f(uv)f(u/v)$. Obviously it is analytic as a function of $v$ in $\Re e\, v > 0$. However, it is also analytic as a function of $w = v + v^{-1}$ for $\Re e\, w > 0$. To see that, consider the mapping $v \to \frac{v-1}{v+1} = z$. It maps the open right half–plane on the open unit disc and has the inverse $v = \frac{1+z}{1-z}$. Now $g$, as a function of $z$, is invariant with respect to the transformation $z \to -z$. Hence it is an analytic function of $z^2$ and therefore also analytic as a function of $2 \cdot \frac{1+z^2}{1-z^2} = v+v^{-1} = w$. Certainly (4.3.1) holds at least for $\Re e\, w \geq 2$ since the integral is then convergent and represents an analytic function. However, it is convergent also for $\Re e\, w > 0$ for otherwise, by a standard result for the Laplace transform of a nonnegative measure, there would have been a singularity of the analytic function at some real point $w_0 \in (0,2]$; see, e.g., Widder (1946, p. 58). □

**Proof of Theorem 4.3.1.** In view of Theorem 3.1.3, it suffices to show that the mgf $\varphi$ of $f$ is analytic in $\mathbb{C}\backslash[0,\infty)$ and $\Im m[\varphi'(s)\overline{\varphi(s)}] \geq 0$ for $\Im m\, s > 0$. The defining expression

$$\varphi(s) = \int e^{sx} f(x)dx \tag{4.3.2}$$

shows that $\varphi$ is analytic for $\Re e\, s < 0$. For $\Im m\, s > 0$ another expression must be used. Let $\theta$ be some angle such that $0 \leq \theta < \pi/2$. Then

$$\varphi(s) = e^{i\theta} \int \exp\{se^{i\theta}x\}\, f(e^{i\theta}x)dx \tag{4.3.3}$$

if $\pi/2 - \theta < \arg[s] \leq \pi$ (or $< 3\pi/2 - \theta$), i.e., if $\Re e[se^{i\theta}] < 0$. In fact, for $s < 0$ the integrals in (4.3.2) and (4.3.3) are equal. This follows by an integration of $e^{sz}f(z)$ along the contour given in the following figure.

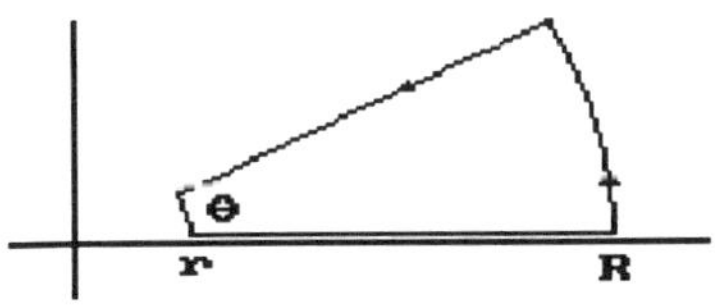


Figure 4.3.1

The conditions (b) and (c) guarantee that the integrals along the arcs tend to zero as $r \to 0$, $R \to \infty$. For $\Im m\, s < 0$, $\varphi(s)$ is defined by (4.3.3) with $\theta$ negative, but we have no use for that expression. In (4.3.3) the choice $\theta = \pi/2$, working for all $s$ in $\Im m\, s > 0$, is the most convenient one but might give a divergent integral.

We now also have, for $\pi/2 - \theta < \arg[s] \le \pi$,

$$\varphi'(s) = (e^{i\theta})^2 \int x \cdot \exp\{se^{i\theta}x\}\, f(e^{i\theta}x)dx.$$

Now

$$J(s) = \varphi'(s)\overline{\varphi(s)} = \iint x \cdot \exp\{i\theta + se^{i\theta}x + \bar{s}e^{-i\theta}y\}\, f(e^{i\theta}x)f(e^{-i\theta}y)dxdy$$

since $\overline{f(e^{i\theta}y)} = f(e^{-i\theta}y)$ by the principle of reflection; $f(x)$, $x > 0$, is real. The double integral is absolutely convergent. We make the hyperbolic substitution $x = uv$, $y = u/v$ and get

$$J(s) = \iint 2u^2 \exp\{i\theta + se^{i\theta}uv + \bar{s}e^{-i\theta}u/v\}\, f(ue^{i\theta}v)f(u/(e^{i\theta}v))\, dvdu\,.$$

By the Lemma,

$$f(ue^{i\theta}v)f(u/(e^{i\theta}v)) = \int \exp\{-\lambda u(e^{i\theta}v + e^{-i\theta}/v)\}\, K(d\lambda;u)\,.$$

Using this expression and then interchanging the order of integration between $v$ and $\lambda$, we see that

$$J(s) = \int\int 2u^2\, I(s,u,\lambda)\, K(d\lambda;u)\, du\,,$$

where

$$I(s,u,\lambda) = \int \exp\{i\theta - u(\lambda - s)e^{i\theta}v - u(\lambda-\bar{s})e^{-i\theta}/v\}\, dv.$$

This expression can be simplified in a nice way. The integrand is analytic in $v$ and by a contour integration it is easily seen that the integral can be calculated by an integration along the ray $v = u(\lambda-\bar{s})e^{-i\theta}\cdot\rho$, $\rho > 0$, instead of along $\mathbb{R}_+$. We then get

$$I(s,u,\lambda) = \int u(\lambda-\bar{s})\cdot\exp\{-u^2|\lambda-s|^2\rho - \rho^{-1}\}\, d\rho\,. \tag{4.3.4}$$

The angle $\theta$ has vanished. The imaginary part of $I(s,u,\lambda)$ is positive for $\Im m\, s > 0$, and hence so is the imaginary part of $J(s) = \varphi'(s)\overline{\varphi(s)}$. This ends the proof. □

The pdf's that are HCM are studied more carefully in Chapter 5, where a canonical representation formula is found. Here some important examples are considered. In these examples Conditions (a), (b), and (c) above are trivially satisfied and are not commented upon.

**Example 4.3.1** ***(Beta distribution of the second kind; Ratio of Gamma variables).*** Let

$$f(x) = C\cdot x^{\beta-1}(1+cx)^{-\gamma},\ x > 0,\ (\gamma > \beta > 0, c > 0).$$

Now

$$f(uv)f(u/v) = C^2u^{2\beta-2}\cdot(1 + c^2u^2 + cu(v+v^{-1}))^{-\gamma}$$

and is CM in $w = v + v^{-1}$. The corresponding nonnegative measure $K(d\lambda;u)$ is of the Gamma type. Thus $f$ is HCM and by the Main HCM–theorem, $f \in \mathcal{T}$. The total U–measure equals $\beta$. The result in Example 4.2.1 for the Pareto distribution is generalized. □

**Example 4.3.2** ***(Generalized inverse Gaussian distribution).*** Let

$$f(x) = C\cdot x^{\beta-1}\cdot\exp\{-c_1x - c_2\cdot x^{-1}\},\quad x > 0\quad (c_1,c_2 > 0,\ \beta \in \mathbb{R}).$$

By the Main HCM–theorem, $f \in \mathcal{T}$ since

$$f(uv)f(u/v) = C^2u^{2\beta-2}\cdot\exp\{-(c_1u + c_2\cdot u^{-1})\cdot(v + v^{-1})\}$$

is CM in $w$. The corresponding measure $K(d\lambda;u)$ has all its mass $C^2u^{2(\beta-1)}$ at the point $\lambda_0(u) = c_1 + c_2u^{-2}$. The total U–measure equals $\infty$ since $c_2 > 0$. Letting $c_1 \to 0$, we see that the inverse Gamma distribution is in $\mathcal{T}$. □

**Example 4.3.3** ***(Lognormal distribution).*** Consider

$$f(x) = \frac{1}{\sqrt{2\pi}}\,\sigma^{-1}\cdot x^{-1}\cdot\exp\{-\frac{(\log x)^2}{2\sigma^2}\}$$

The general Lognormal density is obtained by just a change of scale. We get

$$f(uv)f(u/v) = \frac{1}{2\pi}\,\sigma^{-2}\cdot u^{-2}\cdot\exp\{-\frac{(\log u)^2}{\sigma^2} - \frac{(\log v)^2}{\sigma^2}\}.$$

We shall show that this function is CM in $w$. In view of the composition theorem for CM functions it suffices to verify that $(\log v)^2$, which indeed is a function of $w$, has a CM derivative with respect to $w$. For $v \neq 1$, and by continuity also for $v = 1$,

$$\frac{d}{dw}(\log v)^2 = \frac{d}{dv}(\log v)^2/\frac{dw}{dv} = \frac{2\log v}{v}/(1-v^{-2}) = \frac{\log v - \log(v^{-1})}{v - v^{-1}}.$$

Now $\log v$ is a Pick function and, by Example 2.4.1,

$$\log v = \int_{-\infty}^{0}\left(\frac{1}{t-v} - \frac{1}{t-1}\right)dt.$$

Hence

$$\frac{d}{dw}(\log v)^2 = \frac{1}{v-v^{-1}}\int_{-\infty}^{0}\left(\frac{1}{t-v} - \frac{1}{t-v^{-1}}\right)dt = \int_{-\infty}^{0}\frac{1}{1+t^2-tw}\,dt\,.$$

It is obvious that the derivative is CM in $w$. Hence the Lognormal distribution is a GGC. The total U–measure equals $\infty$. The measure $K(d\lambda;u)$ admits no simple expression. □

**Example 4.3.4** ***(Generalized Gamma distribution; Power of Gamma variable).*** Consider

$$f(x) = C\cdot x^{\beta-1}\exp\{-cx^{\alpha}\},\quad x>0\quad (0<|\alpha|\le 1,\ \beta/\alpha>0)$$

corresponding to the power $1/\alpha$ of an rv $X \sim \mathrm{Gamma}(\beta/\alpha, c)$. We have

$$f(uv)f(u/v) = C^2\cdot u^{2(\beta-1)}\exp\{-cu^{\alpha}(v^{\alpha}+v^{-\alpha})\}.$$

Because of symmetry it suffices to consider the case $\alpha>0$. Now $v^{\alpha}$ is a Pick function and we have as is easily verified

$$v^{\alpha} = \int_{-\infty}^{0}\left(\frac{1}{t-v} - \frac{1}{t}\right)\frac{\sin(\alpha\pi)}{\pi}|t|^{\alpha}dt\,.$$

Hence

$$\frac{d}{dw}(v^{\alpha}+v^{-\alpha}) = \frac{\alpha(v^{\alpha} - v^{-\alpha})}{v - v^{-1}} = \frac{\alpha\sin(\alpha\pi)}{\pi}\int_{-\infty}^{0}\frac{1}{1+t^2-tw}|t|^{\alpha}dt\,.$$

Since this derivative is CM in $w$, so is $f(uv)f(u/v)$. Thus the generalized Gamma distribution is a GGC for $|\alpha|\le 1$. The total U–measure equals $\beta$ for $\alpha>0$ and $\infty$ for $\alpha<0$. As the formulae above indicate and as was mentioned in Section 2.2, the Lognormal distribution appears as a limit distribution as $\alpha\to 0$. □

What are the U–measures for the pdf's in the examples above? As the following theorem shows, the U–measure can be expressed in terms of the measures $K(d\lambda;u)$, $u>0$. However, $U$ can seldom be given explicitly.

**Theorem 4.3.2** ***(Representation of U–measure).*** *Let a pdf $f$ satisfy the assumptions in Theorem 4.3.1. Let moreover $K(d\lambda;u)$ have a density $k(\lambda;u)$ being continuous in $\lambda$ $(\lambda > 0)$. Let $\ell(\lambda;\zeta)$ be the Laplace transform of the function $k(\lambda;\sqrt{\xi})$ with respect to $\xi$. Then, under some additional regularity conditions which are not made precise, the U–measure has a density given by*

$$U'(s) = \pi^{-1}\cdot \Im m[\varphi'(s)\overline{\varphi(s)}]/|\varphi(s)|^2\,, \quad s > 0,$$

*where*

$$\Im m[\varphi'(s)\overline{\varphi(s)}] = 2\pi\cdot \int u\cdot k(s;u)du = \pi\cdot \ell(s;0),$$

*and*

$$|\varphi(s)|^2 = \iint \ell(\lambda;(s-\lambda)^2\rho)\,\rho^{-1}e^{-1/\rho}\,d\rho d\lambda\,. \tag{4.3.5}$$

**Proof** (somewhat unstringent). By calculations completely analogous to those in the proof of the Main HCM–theorem we get, for $\Im m\ s > 0$,

$$|\varphi(s)|^2 = \varphi(s)\overline{\varphi(s)} = \iint 2u\left(\int \rho^{-1}\cdot \exp\{-u^2|\lambda-s|^2\rho - \rho^{-1}\}\,d\rho\right) K(d\lambda;u)du\,.$$

Letting then the complex s tend to a real and positive value, substituting $\xi = u^2$, and interchanging the order of integration, we get (4.3.5). Further, from the proof of the Main HCM–theorem, for $\Im m\ s > 0$,

$$\Im m[\varphi'(s)\overline{\varphi(s)}] = \iint 2u^2\,\Im m[I(s,u,\lambda)]\,k(\lambda;u)d\lambda du\,,$$

where

$$\Im m[I(s,u,\lambda)] = u\cdot \Im m\ s\cdot \int \exp\{-u^2|\lambda-s|^2\rho - \rho^{-1}\}\,d\rho\,.$$

The proper interpretation of the limit of $\Im m[I(s,u,\lambda)]$, as a function of $\lambda$, as $\Im m\ s \downarrow 0$ is that it equals $\pi\cdot u^{-1}\delta_0(\lambda-s)$, $s > 0$, where $\delta_0(\cdot)$ is the Dirac function. In fact, the integral of $\Im m[I(s,u,\lambda)]$ with respect to $\lambda$ tends to $\pi/u$ as $\Im m\ s \downarrow 0$ and $\Im m[I(s,u,\lambda)]$ tends to 0 for $\mathrm{Re}\ s \neq \lambda$. By using this limit and interchanging the order of integration, we see that $\Im m[\varphi'(s)\overline{\varphi(s)}] = 2\pi\cdot \int u\cdot k(s;u)du$. The substitution $\xi = u^2$ shows that the last expression equals $\pi\cdot \ell(s;0)$. □

**Example 4.3.5** ***(Beta distribution of the second kind revisited).*** Consider again the pdf $f(x) = C\cdot x^{\beta-1}(1+cx)^{-\gamma}$ in Example 4.3.1. Now, as can be verified,

$$k(\lambda;u) = \frac{C^2u^{2\beta-2}}{c^{\gamma}\Gamma(\gamma)}\,\lambda^{\gamma-1}\exp\{-\lambda(c^{-1} + cu^2)\}\,,$$

Substituting $\xi = u^2$ and then taking the LT, we get

$$\ell(\lambda;\,\zeta) \;=\; \frac{C^2}{c^\gamma}\frac{\Gamma(\beta)}{\Gamma(\gamma)}\,\lambda^{\gamma-1}e^{-\lambda/c}\,(\zeta+\lambda c)^{-\beta}\,.$$

Thus by Theorem 4.3.2, for $s > 0$,

$$\Im m[\varphi'(s)\overline{\varphi(s)}] \;=\; \pi\cdot\ell(s;\,0) \;=\; \pi\cdot\frac{C^2}{c^{\beta+\gamma}}\frac{\Gamma(\beta)}{\Gamma(\gamma)}\,s^{\gamma-\beta-1}e^{-s/c}\,.$$

This generalizes the result in Example 4.2.1; note that $\gamma+1$ and $1/\lambda$ there are $\gamma$ and $c$ here. Further, by (4.3.5) and the substitution $y = 1/\rho$

$$|\varphi(s)|^2 \;=\; \frac{C^2}{c^\gamma}\frac{\Gamma(\beta)}{\Gamma(\gamma)}\int\!\!\int \frac{\lambda^{\gamma-1}y^{\beta-1}e^{-\lambda/c}e^{-y}}{((s-\lambda)^2\,+\,y\lambda c)^\beta}\,dyd\lambda\,.$$

No essential simplification is possible although $|\varphi(s)|^2$ can also be expressed in terms of the confluent hypergeometric function; see Goovaerts, D'Hooge & de Pril (1978) or Ismail & Kelker (1979). □

**Example 4.3.6** ***(Generalized inverse Gaussian distribution revisited).*** Consider the pdf $C\cdot x^{\beta-1}\exp\{-c_1x - c_2x^{-1}\}$, here denoted $f(x;\,\beta,c_1,c_2)$. For $\beta > 0$, the convolution

$$\int_0^x f(x-y;\,\beta,c_1,0)\cdot f(y;\,-\beta,c_1,c_2)\,dy$$

equals $f(x;\,\beta,c_1,c_2)$ as the substitution $y = x/z$ easily shows. Hence, for $\beta > 0$, the U–measure corresponding to $f(x;\,\beta,c_1,c_2)$ equals the the U–measure corresponding to $f(x;\,-\beta,c_1,c_2)$ except for an atom of size $\beta$ at $t = c_1$ of the first U–measure. We may therefore restrict attention to the case $\beta \le 0$. Now, for $\beta \le 0$, $K(d\lambda;u)$ has all its mass $M(u) = C^2u^{2(\beta-1)}$ at $\lambda_0(u) = c_1 + c_2/u^2$. We can not evaluate $\Im m[\varphi'(s)\overline{\varphi(s)}]$ by directly applying Theorem 4.3.2. However, pretending that $K(d\lambda;u)$ has the density $M(u)\cdot g_n(\lambda - \lambda_0(u))$, where $\{g_n\}_1^\infty$ is a sequence of pdf's on $(0,\infty)$ with the one–point distribution at $0$ as limit, we get after some calculation that we should have

$$\ell(\lambda;\zeta) \;=\; \lim_{n\to\infty}\int e^{-\zeta\xi}\,M(\sqrt{\xi})\,g_n(\lambda-\lambda_0(\sqrt{\xi}))\,d\xi \;=\; C^2c_2^\beta\,(\lambda-c_1)^{-\beta-1}\,\exp\{-\zeta\cdot\tfrac{c_2}{\lambda-c_1}\}$$

for $\lambda > c_1$. Since $\varphi(s)$ is real for $s < c_1$, it follows that

$$\Im m[\varphi'(s)\overline{\varphi(s)}] \;=\; \pi\cdot\ell(s;\,0) \;=\; \begin{cases} 0 & \text{for } s < c_1 \\ \pi C^2c_2^\beta\cdot(s-c_1)^{-\beta-1} & \text{for } s > c_1 \end{cases}$$

A less explicit expression for $|\varphi(s)|^2$ is also obtained by (4.3.5). It should be added that $\varphi(s)$ is analytic in $\mathbb{C}\setminus[0,\infty)$ and smooth up to and on the cut except at $s = c_1$. In fact, for any $\beta \in \mathbb{R}$, $\varphi(s) \propto (c_1-s)^{-\beta/2} K_\beta(2\sqrt{c_2}\sqrt{c_1-s})$, where $K_\beta$ is a modified Bessel function of the third kind; see, e.g., Erdélyi et al. (1954, p. 146, Formula 29). □

In most cases the U–measure has to be calculated by numerical calculation of $\arg[\varphi(s)]$ for $s > 0$. In the proof of Theorem 4.3.1 we noticed that, for $0 \le \theta < \pi/2$,

$$\varphi(s) = e^{i\theta} \int \exp\{se^{i\theta}x\}f(e^{i\theta}x)dx\,, \quad \pi/2 - \theta < \arg[s] \le \pi,$$

provided that $f(z)$ is analytic for $\Re e\, z > 0$ and satisfies some other mild conditions. Under slightly stronger conditions we can even set $\theta = \pi/2$ and permit $s$ to be positive (and on the upper side of the cut $[0,\infty)$) in which case

$$\varphi(s) = i\cdot\int e^{isx}f(ix)dx\,, \quad s > 0.$$

If $f(z)$ is analytic in $\mathbb{C}\setminus(-\infty,0]$ and smooth up to and on the cut $(-\infty,0]$ and $|f(z)|$ does not grow too fast as $|z| \to \infty$, we may even use the value $\theta = \pi$ yielding

$$\varphi(s) = -\int_0^\infty e^{-sx}f(-x)dx\,, \quad s > 0, \tag{4.3.6}$$

where $-x$ is supposed to be on the upper side of the cut. To derive (4.3.6) it is simplest to start from the relation $\varphi(s) = \int e^{sx}f(x)dx$, $s < 0$. The transformation $y = -sx$ $(y > 0)$ yields $\varphi(s) = \int e^{-y}f(-y/s)(-s)^{-1}dy$. We see that $\varphi$ is analytic in $\mathbb{C}\setminus[0,\infty)$ and smooth up to and on the cut. The transformation $y = sx$ for $s > 0$ then gives (4.3.6).

**Example 4.3.7** ***(Generalized Gamma distribution).*** Consider $f(x) = C\cdot x^{\beta-1}\exp\{-x^\alpha\}$, $x > 0$ $(0 < \alpha < 1, \beta > 0)$. By a series expansion,

$$f(-x) = C\cdot \sum_{k=0}^{\infty} \frac{(-1)^k}{k!}(-x)^{\beta+k\alpha-1} = -C\cdot \sum_{k=0}^{\infty} \frac{(-1)^k}{k!} x^{\beta+k\alpha-1}\cdot\exp\{i\pi(\beta+k\alpha)\}$$

and hence, by (4.3.6),

$$\varphi(s) = C\cdot s^{-\beta} \sum_{k=0}^{\infty} \frac{(-1)^k}{k!}\Gamma(\beta+k\alpha)\exp\{i\pi(\beta+k\alpha)\}\, s^{-k\alpha}, \quad s > 0.$$

By calculation of $\tan(\arg[\varphi(s)]) = \Im m[\varphi(s)]/\Re e[\varphi(s)]$, one can verify numerically that $\arg[\varphi(s)]$ is increasing and get numerical values for $U(t) = \pi^{-1}\cdot\arg[\varphi(t)]$. Such calculations have given the empirical base for many results only later proved true; cf. Thorin (1977b) and B1978. □

### 4.4 A Multiplicative Property of GGC's

This section, treating size–biased sampling from a GGC, is not central.

Let $f \in \mathcal{T}$ and let $\gamma > 0$ be such that $\int x^{-\gamma} f(x)dx < \infty$. Does the pdf $g(x) = C \cdot x^{-\gamma} f(x)$ always belong to $\mathcal{T}$? The result in Example 3.2.7 indicates that the answer is affirmative. A partial answer is given in this section.

Since

$$x^{-\gamma} = \int \frac{t^{\gamma-1}}{\Gamma(\gamma)} e^{-xt} dt ,$$

it follows that the mgf of $g(x)$ is given by the fractional integral

$$\psi(s) = \frac{C}{\Gamma(\gamma)} \int_{-\infty}^{s} (s-t)^{\gamma-1} \varphi(t)dt , \quad s < 0, \tag{4.4.1}$$

where $\varphi$ corresponds to f. Certainly $\psi$ can be extended to an analytic function in $\mathbb{C}\backslash[0,\infty)$; we just integrate from $-\infty$ to $s \in \mathbb{C}\backslash[0,\infty)$ along any path that does not intersect the cut. The case $\gamma \in \mathbb{N}_0$ is simplest to handle and the following theorem holds.

**Theorem 4.4.1.** *If* $f \in \mathcal{T}$, *and* $x^{-\gamma} f(x)$, $\gamma \in \mathbb{N}_0$, *is integrable, then* $g(x) = C \cdot x^{-\gamma} f(x) \in \mathcal{T}$.

**Proof.** Obviously it suffices to consider the case $\gamma = 1$. In view of the Closure Theorem we may also without loss of generality assume that f is a convolution of n Gamma densities with *different* parameters $t_j$ and shape parameters $u_j$, $u_j < 1$, $j = 1, ..., n$. Hence

$$\varphi(s) = \prod_{j=1}^{n} \left(\frac{t_j}{t_j - s}\right)^{u_j} , \qquad \Sigma u_j > 1.$$

The mgf $\psi$ of g is given by $\int_{-\infty}^{s} \varphi(t)dt$, $s \in \mathbb{C}\backslash[0,\infty)$; for simplicity the constant is forgotten here and in the sequel. To reach an $s > 0$ on the upper side of the cut, we may integrate along the upper side of the real axis. Since $u_j < 1$, the singularities of $\varphi$ cause no problems. We shall apply Theorem 3.1.4 (the Inversion Theorem) and its Remark 1 on the mgf $\psi$. Since $\psi'(s) = \varphi(s)$, we have

$$\psi'(s) \sim C_0 \cdot (-s)^{-\Sigma u_j} \quad \text{and} \quad \psi(s) \sim \frac{C_0}{\Sigma u_j - 1} (-s)^{-\Sigma u_j + 1}, \quad \text{as } |s| \to \infty .$$

These asymptotic relations hold uniformly in $\mathbb{C}\backslash[0,\infty)$. It follows that $\psi'(s)/\psi(s) \to 0$ uniformly as $|s| \to \infty$ (i.e., Condition (b) in Theorem 3.1.4 holds). Further, it follows that the change of the argument of $\psi(s)$ as s moves from $+\infty$ to $-\infty$ along a very large half–circle in the upper half–plane equals $\Delta = -\pi \cdot (\Sigma u_j - 1)$. Consider now the path $\mathbb{R} \ni s \to \psi(s)$ (with s on the upper side of the cut). It consists of broken lines with breaks at $s = t_j$, $j = 1, ..., n$. It always *turns to the left*. We hope it has an appearance of the type displayed in Figure 4.4.1. In this case $\arg[\psi(+\infty)] = \pi \cdot \Sigma u_j - \pi$ and $\arg[\psi(s)]$, $s > 0$, is increasing.

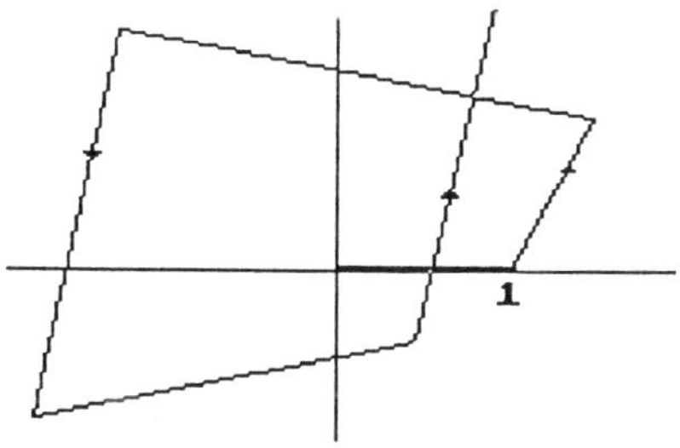


Figure 4.4.1

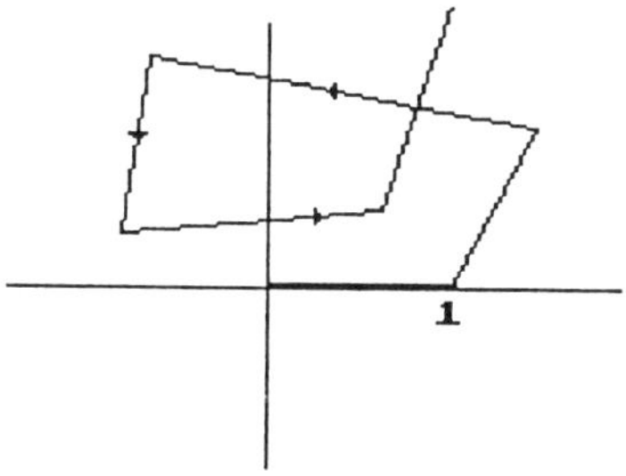


Figure 4.4.2

However, there is also the possibility that the path looks like the path displayed in Figure 4.4.2. There may be loops not moving around the origin and $\psi(s)$ may have some zero $s_0$ (of multiplicity 1) on the cut; that zero is then avoided by performing the integration $\int_{-\infty}^{s} \varphi(t)dt$ along a small half–circle in the upper half–plane around $s_0$. In this case

$$\arg[\psi(+\infty)] = \pi \cdot \Sigma u_j - \pi - 2\pi \cdot L ,$$

where L is the number of loops plus the number of real zeros. If $L = 0$, then $\arg[\psi(s)]$, $s > 0$, is increasing and $\psi(s)$ has no real zeros.

Let now M be the number of zeros of $\psi$ for $\Im m\, s > 0$. By the principle of argument,

$$2\pi \cdot M = \arg[\psi(+\infty)] + \Delta ,$$

with $\Delta$ as defined above. Inserting now the values of $\Delta$ and $\arg[\psi(+\infty)]$ and cancelling terms, we get

$$2\pi \cdot M = -2\pi \cdot L .$$

Hence $M = L = 0$. Thus all the conditions in Theorem 3.1.4 and its Remark 1 hold. The proof is completed. □

Unfortunately the proof does not work to prove that Theorem 4.4.1 holds for all $\gamma > 0$. The property of the path of $\psi(s)$ of turning to the left does not always hold if $\gamma < 1$. To remove the restriction on $\gamma$, a different proof is needed. The proof here has the advantage that it provides a simple geometric interpretation of how the U–measure is transformed when f is multiplied by $x^{-1}$. As the proof also shows, the total of the new U–measure equals the old total minus 1 (if the left–extremity is zero); this is in agreement with Theorem 4.1.4. There are two simple consequences of the theorem:

**Corollary.** *Let $\varphi(s)$ be the mgf of a probability distribution $F$ on $\mathbb{R}_+$. If $\varphi$ is analytic in $\mathbb{C}\setminus[0,\infty)$ and, for some $k \geq 0$, $\Im m[\varphi^{(k+1)}(s)\cdot\overline{\varphi^{(k)}(s)}] \geq 0$ in $\Im m\, s > 0$, then $\varphi$ is the mgf of a GGC and $\Im m[\varphi^{(j+1)}(s)\cdot\overline{\varphi^{(j)}(s)}] \geq 0$ in $\Im m\, s > 0$ for $j = 0, 1, \ldots, k$.*

**Proof.** By the assumption and Theorem 3.1.3, $C\cdot x^k e^{-\epsilon x}F(dx)$ is a GGC for any $\epsilon > 0$. Hence so is $C\cdot x^j e^{-\epsilon x}F(dx)$ for $j = 0, 1, \ldots, k$, by Theorem 4.4.1. □

**Remark.** For $s > 0$, $\Im m[\varphi''(s)\overline{\varphi'(s)}] \geq 0$ means precisely that the path $\mathbb{R} \ni s \to \varphi(s)$ turns to the left. In fact, for a small $\delta > 0$, $\Im m[\varphi''(s)\overline{\varphi'(s)}] \approx \delta^{-1}\cdot\Im m[\varphi'(s+\delta)\overline{\varphi'(s)}]$ and $\Im m[\varphi'(s+\delta)\overline{\varphi'(s)}]$ is the scalar product of the vectors $\varphi'(s+\delta)$ and $i\varphi'(s)$ (= the tangent vector $\varphi'(s)$ rotated $90^\circ$ in the positive direction). □

The second consequence of Theorem 4.4.1 generalizes the theorem.

**Theorem 4.4.2.** *If $f \in \mathcal{T}$, then $C\cdot(1+cx)^{-\gamma}f(x)$ also belongs to $\mathcal{T}$ $(c > 0, \gamma \in \mathbb{N}_0)$.*

**Proof.** By a change of scale, we may assume that $c = 1$. Consider the pdf

$$g(x) = \begin{cases} f(x-1) & \text{for } x > 1 \\ 0 & \text{for } x < 1 \end{cases},$$

i.e., the original GGC translated one unit to the right. Since g belongs to $\mathcal{T}$, so does $C\cdot x^{-\gamma}g(x)$. Translating then this pdf, which has left–extremity 1, one unit to the left, the pdf $C\cdot(1+x)^{-\gamma}f(x)$ is obtained. It must also belong to $\mathcal{T}$. □

## 4.5 Complementary Results and Notes

I (***More convolutions as mixtures***). Theorem 4.1.1 can easily be generalized to be valid also for convolutions of *generalized Gamma distributions*. Let $X_j \sim \text{Gamma}(\beta_j,\cdot)$, $j = 1, 2, \ldots, n$, be independent rv's and set $\beta = \Sigma\,\beta_j$. Then, for any $q \neq 0$, $\Sigma\, X_j^q \overset{d}{=} Y^q\cdot Z$, where $Z > 0$ and $Y \sim \text{Gamma}(\beta,1)$ are independent. In particular, by letting $q \to \infty$, we see that a convolution of Lognormal distributions is a mixture of Lognormal distributions (with $\mu = 0$, in fact). Moreover, for $q = -1$, the result leads to a short proof of Ruben's (1960) result that a convolution of $t(f_j)$–distributions, $j = 1, 2, \ldots, n$, is a scale mixture of $t(f)$–distributions $(f = \Sigma\, f_j)$.

II (***Logconcavity of the cdf of a GGC***). The cdf F of a GGC is logconcave, i.e., $f(x)/F(x)$ is decreasing. In fact, as is easily verified, the cdf of a Gamma distribution is logconcave. Since the class of distributions with logconcave cdf's is closed with respect to convolution (see Barlow & Proschan (1965, p. 38)), and weak limits, the result follows.

III (***Positivity of the density of a GGC***). The pdf of a GGC with $a = 0$ has no zeros in $(0, \infty)$. This can be shown in many ways. For instance, it follows from the fact that any GGC with $a = 0$ is a convolution of an MED and another GGC. More generally, the pdf of an ID distribution on $(0, \infty)$ with left–extremity 0 has no zeros in $(0,\infty)$; Steutel (1971).

IV (***Generalized Gamma distribution***). Consider the pdf $f(x) = C\cdot x^{\beta-1}\exp(-x^{\alpha})$, $x > 0$, for $\alpha > 1$. Then $f \notin \mathcal{T}$ since f is not ID. In fact, $f(x)$ tends more rapidly to 0 as $|x| \to \infty$ than is possible for a non–Normal ID distribution; cf. Ruegg (1970) or Lukacs (1983, pp. 69–70). It is unknown whether or not f is ID for $\alpha < -1$.

**Notes.** Theorems 4.1.1 — 4.1.5 can essentially be found in B1979c and B1984 but the proofs are new. Theorem 4.2.1 appeared in B1981. Example 4.2.1 is due to Thorin (1977a) who showed in a different way that the Pareto distribution is a GGC. Another proof was given by Berg (1981).

Olof Thorin (1912 – )
(Photo from 1976.)

Thorin (1977b, 1978b) found that the Lognormal and the generalized Gamma distributions are GGC's. Thorin's proofs are based on the Inversion Theorem and are ingenious. However, they are very long and technical. In the Lognormal case Thorin proved and used that, for any $c > 0$, $\exp\{-c\cdot(\log(t+1+\sqrt{t^2+2t}))^2\}$ is a completely monotone function of t. In fact, this means that the Lognormal density is HCM. The author, B1979a, found Thorin's technique to work also for a large class of distributions and very roughly Theorem 4.3.1 (Main HCM–theorem) was obtained. The short proof given here was found only recently, B1990.

The result in Example 4.3.1 (Beta distribution) is due to Goovaerts, D'Hooge & de Pril (1978) and Ismail & Kelker (1979). They used properties of special functions in their derivations. The result in Example 4.3.2 is due to Barndorff–Nielsen & Halgreen (1977) but is based on Grosswald's (1976) early result that the inverse Gamma distribution is ID. The possible ID of that distribution was stated as an open problem by Ismail & Kelker (1976). Its ID is implicit in a very early paper by Hammersley (1960); cf. Pitman & Yor (1981).

Theorem 4.3.2 is new and is a partial answer to the often asked question: "What is the U–measure of the Lognormal distribution?". Theorem 4.4.2 appeared in B1988a with a more complicated proof.

# 5. HYPERBOLICALLY COMPLETELY MONOTONE DENSITIES

In this chapter many important consequences of Theorem 4.3.1 are revealed. Moreover, a canonical representation formula for pdf's satisfying the HCM–condition is found.

In Section 5.1 the class $\mathscr{C}$ of pdf's that are HCM is introduced and studied. A subclass $\mathscr{B}$ of $\mathscr{C}$ is the focus of Section 5.2. The pdf's in the subclass admit a canonical representation. In Section 5.3 it is shown that the class $\mathscr{B}$ is surprisingly large. Sections 5.4–5.7 treat related topics and prepare for Chapter 6.

## 5.1 The Class $\mathscr{C}$

We recall that a *function* f on $(0,\infty)$, with values in $\mathbb{R}_+$, is said to be HCM if, for every $u > 0$, $f(uv)f(u/v)$ is CM in $w = v + v^{-1}$. Equivalently, by Bernstein's theorem, f is HCM if

(5.1.1) $$f(uv)f(u/v) = \int_{[0,\infty)}\exp\{-\lambda u(v+v^{-1})\}K(d\lambda;u), \qquad K(d\lambda;u) \geq 0.$$

We let $\mathscr{C}$ denote the class of *pdf's* which are HCM. The regularity conditions in Section 4.3 are here abandoned. This class is studied in this section. However, we start by first looking at all HCM–functions. Examples of such functions are provided by $x^{\beta}$, $\beta \in \mathbb{R}$, $e^{-cx}$, $e^{-c/x}$, and $(1+cx)^{-\gamma}$, $c > 0$, $\gamma > 0$. Important properties of these functions are now listed. Assuming that the functions $f, f_1, f_2, \ldots$ are HCM and measurable, we have:

*i)* *The function $f(cx)$, $c > 0$, is HCM.*

*ii)* *The product $f_1 \cdot f_2 \cdot \ldots \cdot f_n$ is HCM.*

*iii)* *If, for all $x > 0$, $f_n(x) \to g(x)$ as $n \to \infty$, then $g$ is HCM.*

*iv)* *The functions $f(x^{\alpha})$ and hence $x^{\beta}f(x^{\alpha})$ are HCM for $|\alpha| \leq 1$ and $\beta \in \mathbb{R}$.*

*v)* *$f$ is strictly positive (unless $f(x) \equiv 0$) and the functions $f(e^y)$ and $e^y f(e^y)$ are logconcave.*

*vi)* *Constants $B \geq 0$ and $c \in \mathbb{R}$ (depending on $f$) exist such that $f(x) \leq B \cdot x^c$.*

*vii)* *The function $x \to \int_0^{\infty} f_1(xy)f_2(y)dy$ is HCM (when finite for all $x$). In particular, the LT of $f_2$ is HCM.*

*viii* *$f^p(x)$ is HCM for any $p > 0$.*

*ix)* *$f$ can be extended to an analytic function in $\mathbb{C}\backslash(-\infty, 0]$.*

*x)* *$f(0+) > 0$ iff $f$ is decreasing. In this case $f$ is CM.*

*xi)* *If $f$ is decreasing, then the functions $f(x+\delta)$, $\delta \geq 0$, and $x \to \int_x^{\infty} (y-x)^{\gamma-1}f(y)dy$, $\gamma > 0$, are HCM.*

Here (i) is obvious while (ii) and (iii) are rather immediate consequences of the fact that complete monotonicity is preserved under multiplication and pointwise limits. The properties (iv) – (vi) are verified below. The property (vii) will be apparent later on in this section. Several variants of (vii) appear when it is used in conjunction with (iv). The properties (viii) and (ix) are hard to prove. Proofs are given only in Section 5.3. The properties (x) and (xi) will be established in Sections 5.4 and 6.2.

To prove (iv), we set $g(x) = f(x^\alpha)$ and recall (5.1.1). Hence

$$g(uv)g(u/v) = \int \exp\{-\lambda u^\alpha(v^\alpha+v^{-\alpha})\}\ K(d\lambda;u)\ .$$

Now (iv) is obvious from the fact that $v^\alpha + v^{-\alpha}$ has a CM derivative with respect to $w = v + v^{-1}$; cf. Example 4.3.4.

Turning to (v), we note that if $f(u) > 0$ for some $u$, then $K(d\lambda;u) \not\equiv 0$ by (5.1.1) for $v = 1$. Hence $f(uv)f(u/v) > 0$ for all $v$. Thus $f$ is strictly positive unless $f \equiv 0$. Further, (5.1.1) gives for all $u$, $0 < f(uv)f(u/v) \le (f(u))^2$, i.e.,

$$0 < f(e^{y+h})f(e^{y-h}) \le (f(e^y))^2.$$

Hence the function $f(e^y)$ is logconcave and so is $e^y f(e^y)$. (For this conclusion to be valid some extremely weak regularity condition on $f$ is needed. It suffices that $\log f$ is bounded in some interval; cf., e.g. Boas (1981, p. 164). Measurability should also suffice.) It follows that $f$ is continuous and differentiable at least a.e.

Finally, (vi) is a simple consequence of the fact that a logconcave function is below each of its tangents.

We now turn to the class $\mathscr{C}$ of HCM pdf's. Property (v) above shows that if $X \sim \mathscr{C}$, then $\log X$ has a logconcave pdf. By (iv) we get:

*If $X$ is an rv and $X \sim \mathscr{C}$, then $X^q \sim \mathscr{C}$ for all $q \in \mathbb{R}$ such that $|q| \ge 1$.*

In fact, the pdf of $X^q$ equals $f_{X^q}(x) = |\alpha|\cdot x^{\alpha-1} f_X(x^\alpha)$ with $\alpha = q^{-1}$ and it is HCM by (iv). It is worth noticing that hence $X \sim \mathscr{C}$ *iff* $1/X \sim \mathscr{C}$. It is easy to verify that the convolution of two different Exponential distributions is not in $\mathscr{C}$. Thus $\mathscr{C}$ is not closed with respect to convolution. However, $\mathscr{C}$ has another important property which is formulated as a theorem.

**Theorem 5.1.1** ***(Multiplication theorem).*** *If $X \sim \mathscr{C}$ and $Y \sim \mathscr{C}$, are independent rv's, then $X \cdot Y \sim \mathscr{C}$ and $X/Y \sim \mathscr{C}$.*

**Proof.** Obviously it suffices to show that $X/Y \sim \mathscr{C}$. The pdf of $X/Y$ equals $h(x) = \int f(xy)g(y)dy$, where here $f = f_X$ and $g(y) = yf_Y(y)$. Note that $g$ is also HCM. We have to show that, for each $s > 0$,

$$J = h(st)h(s/t) = \iint f(stx)f(\tfrac{s}{t}\cdot y)\, g(x)g(y)\, dxdy$$

is CM in $t + t^{-1}$. Making now the *hyperbolic substitution* $x = uv$, $y = u/v$ with Jacobian $-2u/v$, we get

$$J = \int 2u \int \frac{1}{v} f(su\cdot tv)f(su/(tv))\, g(uv)g(u/v)\, dv\, du\,.$$

It suffices to show that, for fixed $s$ and $u$, the interior v–integral is CM in $t + t^{-1}$. By the assumptions,

$$f(su\cdot tv)f(su/(tv)) = \int \exp\{-\lambda su(tv + \tfrac{1}{tv})\}\, K_f(d\lambda;su)$$

$$g(uv)g(u/v) = \int \exp\{-\lambda' u(v + \tfrac{1}{v})\}\, K_g(d\lambda';u)\,,$$

where the K–measures are nonnegative. Interchanging the order of integration between $v$ and $(\lambda,\lambda')$ and setting $a = \lambda su$, $b = \lambda' u$, we see that it suffices to check that

$$\int \frac{1}{v}\cdot \exp\{-a(tv + \tfrac{1}{tv}) - b(v + \tfrac{1}{v})\}\, dv$$

is CM in $t + t^{-1}$ (for fixed $a$ and $b$). Rewriting the exponent as $-(at+b)v - (\frac{a}{t} + b)\frac{1}{v}$ and then substituting $\rho = v/(\frac{a}{t} + b)$, we find that the last integral equals

$$\int \frac{1}{\rho}\cdot \exp\{-(a^2 + b^2 + ab(t+t^{-1}))\cdot\rho - \tfrac{1}{\rho}\}\, d\rho\,.$$

The desired complete monotonicity is now obvious. □

Property (vii) for the HCM–functions is apparent from the proof above. The class $\mathscr{C}$ has also the following notable property:

*Let $X \sim \mathscr{C}$ and $Y \sim \mathscr{C}$ be independent rv's. Then the conditional pdf of $V = \sqrt{X/Y}$ given $U = \sqrt{X\cdot Y} = u$ belongs to $\mathscr{C}$.*

In fact, $f_{V|U=u}(v) \propto \frac{2u}{v} f_X(uv)\cdot f_Y(u/v)$ and is HCM by (i), (ii), and (iv). The conditional pdf of $V' = X/Y$ given $U' = X\cdot Y = u$ is proportional to

$\frac{1}{2v} f_X(\sqrt{uv}) \cdot f_Y(\sqrt{u/v})$ and belongs to $\mathscr{C}$ as well.

Theorem 4.3.1 states that if f belongs to $\mathscr{C}$ and satisfies some regularity conditions, then $f \in \mathscr{T}$. In fact, these conditions are superfluous. This is a main theorem of this section.

**Theorem 5.1.2** ***(Main HCM–theorem, improved version).*** $\mathscr{C} \subset \mathscr{T}$.

**Proof.** Let $f \in \mathscr{C}$ and let X be an rv with this pdf. Let further, for $n \in \mathbb{N}$, $Y_n$ be an independent Gamma(n, n)–variable. Consider $Z_n = Y_n \cdot X$. By Theorem 5.1.1, $Z_n \sim \mathscr{C}$. As $Z_n = Y_n/X^{-1}$, its pdf is given by

$$f_{Z_n}(z) = C \cdot z^{n-1} \int e^{-znx} x^n f_{X^{-1}}(x)\, dx \qquad (C = n^n/\Gamma(n)),$$

which can be easily verified to satisfy the conditions (a), (b), and (c) of Theorem 4.3.1. Hence $Y_n \cdot X \sim \mathscr{T}$. Since $Y_n \cdot X$ tends in distribution to X as $n \to \infty$ and $\mathscr{T}$ is closed with respect to weak limits, it follows that $f \in \mathscr{T}$. □

In view of Theorem 4.1.3, Theorem 5.1.2 shows in particular that a pdf in $\mathscr{C}$ can be differentiated any number of times on $(0, \infty)$. Hence an HCM–function f has the same property since, by (ii) and (vi) above, $C \cdot x^{-c} e^{-\epsilon x} f(x)$ belongs to $\mathscr{C}$ for some $c \in \mathbb{R}$ and any $\epsilon > 0$.

There is also a closure theorem for $\mathscr{C}$:

**Theorem 5.1.3.** *The class $\mathscr{C}$ is closed with respect to weak non–degenerate limits. More precisely: If $f_n \in \mathscr{C}$, $n = 1, 2, \ldots$, converges weakly to $f$ ($\in \mathscr{T}$), then $f \in \mathscr{C}$ and $f_n(x) \to f(x)$ for all $x > 0$.*

**Proof.** By Theorem 4.1.5, $f_n(x) \to f(x)$ for all $x \neq a$, where $a$ is the left–extremity of f. Hence $f \in \mathscr{C}$ if $a = 0$. Thus it suffices to show that $f(x) > 0$ for all $x > 0$. Let $u > a$. Then $f(u) > 0$; cf. Section 4.5. Using that

$$f_n(uv) f_n(u/v) = \int_{[0,\infty)} \exp\{-\lambda u(v+v^{-1})\}\, K_n(d\lambda;u),$$

we see that $\int \exp\{-2\lambda u\} K_n(d\lambda;u) = f_n(u) f_n(u) \to (f(u))^2 > 0$. Hence the sequence $\tilde{K}_n(d\lambda;u) = \exp\{-2\lambda u\} K_n(d\lambda;u)$, $n = 1, 2, \ldots$, has a subsequence converging vaguely to a nonnegative measure $\tilde{K}(d\lambda;u)$ $(\neq 0)$ on $[0, \infty]$. Setting $K(d\lambda;u) = \exp\{2\lambda u\}\tilde{K}(d\lambda;u)$ on $[0, \infty)$, we then have, for all $v > 1$,

$$f_n(uv)f_n(u/v) \to \int_{[0,\infty)} \exp\{-\lambda u(v+v^{-1})\}\, K(d\lambda;u) > 0$$

as n runs through some sequence of integers; note that $v + v^{-1} > 2$. The left hand side tends also to $f(uv)f(u/v)$ for $uv \neq a$ and $u/v \neq a$. Varying v, we see that $f(x) > 0$ for $x > 0$. □

The results in Sections 5.2 and 5.3 will confirm that $\mathscr{C}$ is closed under weak limits. We end this section by deriving some simple properties of the pdf's in $\mathscr{C}$ from the very definition of $\mathscr{C}$. These properties have been derived earlier for the pdf's in $\mathscr{T}$.

Let $f \in \mathscr{C}$. From the logconcavity of $f(e^y)$ it follows that $c(x) = -xf'(x)/f(x)$ is an increasing function. We have

$$f(x) = f(1)\cdot\exp\{-\int_1^x \frac{c(y)}{y}\, dy\}\ . \tag{5.1.2}$$

Let $c_0 = \lim_{x\to 0} c(x)$ and $c_\infty = \lim_{x\to\infty} c(x)$. As f is integrable, it follows that $-\infty \leq c_0 < 1$ and $1 < c_\infty \leq \infty$. If $c_0 \geq 0$, then f is decreasing; otherwise f is first increasing (and $f(0+) = 0$) and then decreasing. In both cases f is unimodal as it should be since $f \in \mathscr{T}$. From (5.1.2) it also follows that

$$1-c_0 = \sup\{\alpha\ ;\ \lim_{x\to 0} f(x)/x^{\alpha-1} = 0\}.$$

In view of Theorem 4.1.4, $1-c_0$ can be identified as the total U–measure for $f \in \mathscr{T}$. Assuming that $c_0 > -\infty$, we easily get from (5.1.2)

$$f(tx)/f(t) \to x^{-c_0} \quad \text{as } t \to 0,$$

i.e., f is regularly varying at $x = 0$. Analogously, if $c_\infty < \infty$, then $f(tx)/f(t) \to x^{-c_\infty}$ as $t \to \infty$. This regular variation at $x = \infty$ is a consequence of the regular variation at $x = 0$ since $X \sim \mathscr{C}$ iff $1/X \sim \mathscr{C}$. The regular variation of f at $x = 0$ holds for any GGC (with left–extremity 0 and finite U–measure); cf. the proof of Theorem 4.1.4.

## 5.2 The $\mathscr{B}$–class

In this section, in part rather technical, a simple and large subclass $\mathscr{B}$ of $\mathscr{C}$ is studied. The pdf's in $\mathscr{B}$ have explicit expressions.

Obviously $\mathscr{C}$ contains all pdf's of the form:

$$(5.2.1) \qquad f(x) = C\cdot x^{\beta-1}\prod_{i=1}^{n}(y_i + x)^{-\gamma_i}, \quad x > 0,$$

where the parameters are positive and $0 < \beta < \Sigma\gamma_i$, and also all pdf's which are limits of *functions* of this form. In particular, a pdf of the form

$$(5.2.2) \qquad f(x) = C\cdot x^{\beta-1}h(x),$$

where $\beta > 0$ and $h(x)$ is the Laplace transform (LT) of a GGC is in $\mathscr{C}$ and so is $f(x) = C\cdot x^{-\beta-1}h(1/x)$ (since $X \sim \mathscr{C}$ iff $1/X \sim \mathscr{C}$). More generally, $\mathscr{C}$ contains every pdf of the form

$$(5.2.3) \qquad f(x) = C\cdot x^{\beta-1}h_1(x)h_2(1/x),$$

with $\beta \in \mathbb{R}$ and, for $j = 1, 2$,

$$(5.2.4) \qquad h_j(x) = \exp\{-b_jx + \int \log(\tfrac{y+1}{y+x})\Gamma_j(dy)\},$$

where $b_j \geq 0$ and here $\Gamma_j(dy)$ is a nonnegative measure on $(0,\infty)$ satisifying $\int (1+y)^{-1}\Gamma_j(dy) < \infty$. The function $h_j(x)$ is the LT of a widened GGC, cf. Section 3.5. The derivative of $-\log(f(uv)f(u/v))$ with respect to $w = v + v^{-1}$ is CM and this implies that $f$ is HCM. In fact,

$$(5.2.5) \qquad \frac{d}{dw}[-\log(f(uv)f(u/v))] = b_1u + \frac{b_2}{u} + \int \frac{1}{\frac{y}{u}+\frac{u}{y}+w}\,\Gamma_1(dy) + \int \frac{1}{uy+\frac{1}{uy}+w}\,\Gamma_2(dy).$$

The class of pdf's of the form (5.2.3) is denoted $\mathscr{B}$. The pdf's in $\mathscr{B}$ can also be obtained as limits of functions of the form (5.2.1). No other pdf's can appear as such limits as will be established later on.

In (5.2.3), $\beta$, $b_1$, $b_2$, $\Gamma_1$, and $\Gamma_2$ are not uniquely determined. This is perhaps simplest to see by looking at the pdf

$$f(x) = C\cdot x^{\beta-1}\prod_{i=1}^{n}(y_i + x)^{-\gamma_i}.$$

It can be rewritten as

$$f(x) = C'\cdot x^{\beta'-1}\prod_{i;y_i>1}(y_i + x)^{-\gamma_i}\cdot\prod_{i;y_i\leq 1}(\tfrac{1}{y_i} + \tfrac{1}{x})^{-\gamma_i}, \qquad \beta' = \beta - \sum_{i;y_i\leq 1}\gamma_i,$$

where $\beta'$ may be negative. Now $f(x)$ is of the form (5.2.3) and the measures $\Gamma_1$ and $\Gamma_2$ have their masses on $(1,\infty)$ and $[1,\infty)$, respectively. In the same way it follows that *a general pdf of the form (5.2.3) can be rewritten into a form such that the new measures* $\Gamma_1$ *and* $\Gamma_2$ *are concentrated on* $(1,\infty)$ *and* $[1,\infty)$*, respectively.*

In the sequel, this particular representation will be referred to as the ***canonical representation*** although it is not always the most convenient one. However, as will be proved later on, in this representation, $\beta$, $b_1$, $b_2$, $\Gamma_1$, and $\Gamma_2$ are uniquely determined which is an advantage. Moreover, $h_1$ and $h_2$ in (5.2.3) become LT's of GGC's if from each exponent the term $\int \log(\frac{y+1}{y})\Gamma_j(dy)$ is subtracted and transferred to C.

In terms of the parameters in the canonical representation the total U–measure of a pdf in $\mathcal{B}$ ($\subset \mathcal{T}$) is given by

$$\int U(dt) = \beta + \Gamma_2([1,\infty)) + b_2\cdot\infty \qquad (0\cdot\infty = 0) \tag{5.2.6}$$

as easily can be verified by an application of Theorem 4.1.4.

The generalized inverse Gaussian density $f(x) = C\cdot x^{\beta-1}\exp\{-b_1x-b_2x^{-1}\}$ is in $\mathcal{B}$. Obviously $\Gamma_1 = \Gamma_2 \equiv 0$ and the total U–measure is $\infty$ if $b_2 > 0$; otherwise it is $\beta$. Let us now consider two other examples.

Example 5.2.1 ***(Lognormal distribution).*** The pdf of the general Lognormal distribution can be written as

$$f(x) = C\cdot x^{\mu/\sigma^2-1}\cdot\exp\{-\frac{(\log x)^2}{2\sigma^2}\} \qquad (\mu \in \mathbb{R}).$$

To verify that it belongs to $\mathcal{B}$, we must show that

$$\frac{(\log x)^2}{2\sigma^2} = -\int \log(\frac{y+1}{y+x})\Gamma_1(dy) - \int \log(\frac{y+1}{y+x^{-1}})\Gamma_2(dy)$$

for some nonnegative measures $\Gamma_1$ and $\Gamma_2$ on $(1,\infty)$ and $[1,\infty)$, respectively. As equality holds for $x = 1$, it suffices to verify that the derivatives of the two sides are equal, i.e., that

$$\sigma^{-2}\frac{\log x}{x} = \int \frac{1}{x+y}\Gamma_1(dy) - x^{-2}\int \frac{1}{y+x^{-1}}\Gamma_2(dy)\,.$$

For $j = 1, 2$, let $\Gamma_j(dy) = \sigma^{-2}y^{-1}dy$, $y > 1$. By partial fraction decompositions, we find that the right hand side above equals

$$\sigma^{-2}\int_1^\infty \frac{1}{x}\cdot(\frac{1}{y+x^{-1}} - \frac{1}{x+y})\,dy = \sigma^{-2}\frac{\log x}{x}\,.$$

Thus the Lognormal density belongs to $\mathcal{B}$ and the measures in the canonical representation are as given above. Further, $\beta = \mu/\sigma^2$ and $b_1 = b_2 = 0$. □

**Example 5.2.2** ***(Generalized Gamma distribution).*** Let

$$f(x) = C\cdot x^{\beta-1}\cdot\exp\{-x^{\alpha}\},\quad x>0,\ (0<\alpha<1,\ \beta>0).$$

Here $h(x) = \exp\{-x^{\alpha}\}$ is the LT of a GGC and thus $f \in \mathscr{B}$. We have from Example 3.2.1

$$f(x) = C\cdot x^{\beta-1}\cdot\exp\{\int\log(\tfrac{y}{y+x})\,\tfrac{\alpha\sin(\alpha\pi)}{\pi}y^{\alpha-1}dy\}\,.$$

This is the simplest representation of the type (5.2.3), though not the canonical one. □

In connection with Example 5.2.2 it may be added that if in the canonical representation $\Gamma_2([1,\infty)) < \infty$, then $f$ can also be represented as

$$f(x) = C\cdot x^{\beta'-1}\cdot\exp\{-b_1x - \tfrac{b_2}{x} + \int\log(\tfrac{y+1}{y+x})\tilde{\Gamma}_1(dy)\}\,,$$

where $\beta' = \beta + \Gamma_2([1,\infty))$ and $\tilde{\Gamma}_1(dy)$ equals $\Gamma_1(dy)$ on $(1,\infty)$ whereas on $(0,1]$ it equals the measure induced by $\Gamma_2$ and the transformation $y \to y^{-1}$. If $b_2 = 0$, this representation is possible precisely when the total U–measure is finite; cf. (5.2.6).

We now turn to a proof of the uniqueness in the canonical representation, i.e., we shall prove the following theorem.

**Theorem 5.2.1** ***(Uniqueness theorem).*** *If $f \in \mathscr{B}$ and, in (5.2.3), $\Gamma_1$ and $\Gamma_2$ are assumed to be concentrated on $(1,\infty)$ and $[1,\infty)$, respectively, then $C$, $\beta$, $b_1$, $b_2$, $\Gamma_1$, and $\Gamma_2$ are uniquely determined.*

**Proof.** Set $\gamma = \Gamma_2(\{1\})$. Differentiating $\log f(x)$ and $\log f(x^{-1})$, we get

$$\begin{aligned}\psi_1(x) = f'(x)/f(x) &= -b_1 + b_2x^{-2} + \frac{\beta-1}{x} + \frac{\gamma}{x(1+x)}\\ &\quad - \int_{(1,\infty)}\frac{1}{y+x}\Gamma_1(dy) + x^{-2}\int_{(1,\infty)}\frac{1}{y+x^{-1}}\Gamma_2(dy)\end{aligned}$$

and

$$\begin{aligned}\psi_2(x) = -x^{-2}f'(x^{-1})/f(x^{-1}) &= b_1x^{-2} - b_2 - \frac{\beta-1}{x} - \frac{\gamma}{1+x}\\ &\quad + x^{-2}\int_{(1,\infty)}\frac{1}{y+x^{-1}}\Gamma_1(dy) - \int_{(1,\infty)}\frac{1}{y+x}\Gamma_2(dy)\,.\end{aligned}$$

All the functions above are analytic in $\mathbb{C}\setminus(-\infty,0]$ and hence the relations hold for all complex x in that region. Obviously $b_j = -\lim_{x\to+\infty}\psi_j(x)$, $j = 1, 2$, and

$$\gamma = -\lim_{\eta\downarrow 0} i\eta\cdot\psi_1(-1+i\eta) = -\lim_{\eta\downarrow 0} i\eta\cdot\psi_2(-1+i\eta) \qquad (i = \sqrt{-1}).$$

We see that all the terms that give $\psi_j(x)$ except the term $\int_{(1,\infty)}(y+x)^{-1}\,\Gamma_j(dy)$ can be analytically continued across $(-\infty,-1)$ and are real in that interval. The inversion formula (2.4.2) for Pick functions then gives

$$\int_{(c,d)} \Gamma_j(dy) = \lim_{\eta\downarrow 0} \frac{1}{\pi} \int_{(-d,-c)} \Im m[\psi_j(y+i\eta)]\,dy$$

provided that $(c, d) \subset (1,\infty)$ and $\Gamma_j(\{c\}) = \Gamma_j(\{d\}) = 0$. Thus $b_1, b_2, \Gamma_1$, and $\Gamma_2$ are uniquely determined by f and hence so are $\beta$ and C. □

Of course, the given proof provides also a way of finding the explicit canonical representation of an f belonging to $\mathscr{B}$. In particular, if $\pi^{-1}\cdot\Im m[\psi_j(x)]$ is continuous up to and on the cut $(-\infty,-1)$, then $\Gamma_j$ has a continuous density on $(1,\infty)$ given by $\Gamma_j'(y) = \pi^{-1}\cdot\Im m[\psi_j(-y)]$, where $-y$ is on the upper side of the cut. For the standard Lognormal density $(\mu = 0,\ \sigma = 1)$, we have $\psi_j(x) = -x^{-1} - x^{-1}\cdot\log x$, $j = 1, 2$. Since $\pi^{-1}\cdot\Im m[-x^{-1}\cdot\log x] = -x^{-1}$ for $x < -1$, it follows that $\Gamma_j'(y) = y^{-1}$, $y > 1$. In fact, in Example 5.2.1 the measures $\Gamma_j$ were originally found in this way. A given pdf f can always be proved or disproved to belong to $\mathscr{B}$ by this complex technique. First the parameters and measures in the possible canonical representation are found and then it is checked whether the canonical representation really is valid.

The class $\mathscr{B}$ is closed with respect to weak non–degenerate limits. This result is an immediate consequence of Theorem 4.1.5 and Theorem 5.2.2 below which asserts that the $\mathscr{B}$–class is closed with respect to pointwise limits. Theorem 5.2.2 is formulated in terms of a slight modification of the canonical representation for pdf's in $\mathscr{B}$. We have, for $f \in \mathscr{B}$,

$$(5.2.7)\qquad f(x) = C\cdot x^{\beta-1}\cdot\exp\left\{\int_{(1,\infty]}\log\left(\frac{y+1}{y+x}\right) y\nu_1(dy) + \int_{[1,\infty]}\log\left(\frac{y+1}{y+x^{-1}}\right) y\nu_2(dy)\right\},$$

where the nonnegative measures $\nu_1$ on $(1,\infty]$ and $\nu_2$ on $[1,\infty]$ are bounded. In fact, $\nu_j(dy) = y^{-1}\cdot\Gamma_j(dy)$ and $\nu_j(\{\infty\}) = b_j$, $j = 1, 2$.

**Theorem 5.2.2** ***(Closure theorem for $\mathscr{B}$).*** *Let $f_n$, $n = 1, 2, \dots$, be a sequence of pdf's or functions of the form (5.2.7) defined by $C_n$, $\beta_n$, $\nu_{1n}$, and $\nu_{2n}$. If the sequence converges to a strictly positive limit for all $x$ in some open interval, then it converges for all $x > 0$ and the limit $f(x)$ is of the form (5.2.7). If $f$ is defined by $C$, $\beta$, $\nu_1$, and $\nu_2$, then:*

*i)* $C - \lim C_n$

*ii)* *For $j = 1, 2$, $\nu_j$ is a vague limit of $\nu_{jn}$ on $(1,\infty]$*

*iii)* $$\int_{(1,\infty]} \nu_1(dy) + \int_{[1,\infty]} \nu_2(dy) = \lim_{n\to\infty} \Big( \int_{(1,\infty]} \nu_{1n}(dy) + \int_{[1,\infty]} \nu_{2n}(dy) \Big)$$

*iv)* $$\beta - \int_{(1,\infty]} \nu_1(dy) = \lim_{n\to\infty} \Big(\beta_n - \int_{(1,\infty]} \nu_{1n}(dy) \Big)$$

*Conversely, (i) – (iv) guarantee that $f_n(x) \to f(x)$ for all $x > 0$.*

**Proof** (rather technical). Rescaling if necessary, we may suppose that $f_n(x) \to f(x)$ for all $x$ in an open interval $I$ containing the point 1. Hence $C_n = f_n(1)$ converges (towards C). Obviously, for $x \neq 1$ and $x, x^{-1} \in I$,

$$\log\left[\frac{f_n(x)f_n(x^{-1})}{(f_n(1))^2}\right] = \int_{(1,\infty]} g(y,x)\, y\nu_{1n}(dy) + \int_{[1,\infty]} g(y,x)\, y\nu_{2n}(dy),$$

where $g(y,x) = \log\left[\dfrac{y^2 + 2y + 1}{y^2 + (x+x^{-1})y + 1}\right]$, converges towards a finite limit. Clearly $g(y,x) < 0$. Further, the function $y \to y\cdot g(y,x)$ is strictly increasing to the negative limit $2 - (x + x^{-1})$ at $y = \infty$. It follows that the sequences $\int_{(1,\infty]}\nu_{1n}(dy)$ and $\int_{[1,\infty]}\nu_{2n}(dy)$ are bounded. Considering $\nu_{1n}$ as a measure on $[1,\infty]$ with mass 0 at $y = 1$, we may thus thin every subsequence of the sequence $\{\nu_{1n}, \nu_{2n}\}$ to a sequence $\{\nu_{1n_i}, \nu_{2n_i}\}$, $i = 1, 2, \dots$, such that $\nu_{1n_i}$ and $\nu_{2n_i}$ converge vaguely towards measures $\tilde\nu_1$ and $\tilde\nu_2$ on $[1,\infty]$. In particular there is vague convergence on $(1,\infty]$. Since the function $y \to y\cdot\log(\frac{y+1}{y+x})$ is continuous on $[1,\infty]$, it follows that, for all $x > 0$,

$$\log\Big(\frac{f_{n_i}(x)}{f_{n_i}(1)}\Big) - (\beta_{n_i}-1)\log x = \int_{[1,\infty]} \log\Big(\frac{y+1}{y+x}\Big)\, y\nu_{1n_i}(dy) + \int_{[1,\infty]} \log\Big(\frac{y+1}{y+x^{-1}}\Big)\, y\nu_{2n_i}(dy)$$

$$\to \tilde\nu_1(\{1\})\log\Big(\frac{2}{1+x}\Big) + \int_{(1,\infty]} \log\Big(\frac{y+1}{y+x}\Big)\, y\tilde\nu_1(dy) + \int_{[1,\infty]} \log\Big(\frac{y+1}{y+x^{-1}}\Big) y\tilde\nu_2(dy)$$

$$= -\tilde\nu_1(\{1\})\log x + \int_{(1,\infty]} \log\Big(\frac{y+1}{y+x}\Big)\, y\nu_1(dy) + \int_{[1,\infty]} \log\Big(\frac{y+1}{y+x^{-1}}\Big)\, y\nu_2(dy),$$

where, for $j = 1, 2$, $\nu_j(dy) = \tilde{\nu}_j(dy)$ on $(1,\infty]$ and

$$\nu_2(\{1\}) = \tilde{\nu}_2(\{1\}) + \tilde{\nu}_1(\{1\}). \tag{5.2.8}$$

As $\log(f_n(x)/f_n(1))$ has a finite limit for some $x \neq 1$, it follows that also the sequence $\beta_{n_i}$ converges. Set

$$\beta = \lim_{i\to\infty}(\beta_{n_i} - \tilde{\nu}_1(\{1\})). \tag{5.2.9}$$

Hence, for all $x > 0$, $f_{n_i}(x) \to f(x)$, where $f(x)$ is of the form (5.2.7) and definied by $C$, $\beta$, $\nu_1$, and $\nu_2$. Then $f(x)$ is analytic in $\mathbb{C}\backslash(-\infty,0]$ and hence unique since $f_n(x) \to f(x)$, $x \in I$. By Theorem 5.2.1, $C$, $\beta$, $\nu_1$, and $\nu_2$ are uniquely determined. Since (5.2.8) and (5.2.9) correspond to (iii) and (iv) in the formulation of the theorem, we conclude from the uniqueness by a well–known reasoning that (i) – (iv) hold. (This does not mean that the sequences $\beta_n$, $\int_{(1,\infty]} \nu_{1n}(dy)$, and $\int_{(1,\infty]} \nu_{2n}(dy)$ necessarily converge. Mass at points close to 1 may alternate between the measures without affecting the limit of $f_n$.) Using the last part of the theorem that (i) – (iv) imply that $f_n(x) \to f(x)$ for all x, which is rather obvious in view of the above, we conclude that $f_n(x) \to f(x)$ for all $x > 0$. □

Often one can show that a complicated pdf f belongs to $\mathscr{B}$ by verifying that it is a limit of functions of the form (5.2.7).

**Example 5.2.3** ***(Nameless distribution).*** Consider, for $0 < \alpha < 1$,

$$f(x) = C\cdot\exp\{-(c_0 + c_1x + \tfrac{c_2}{x})^\alpha\},\quad x > 0,\qquad (c_1,c_2 > 0,\ c_0 \geq 2\sqrt{c_1c_2}\,)$$

Is f in $\mathscr{B}$? Since $\exp\{-s^\alpha\}$ is the LT of a GGC, we have that, for appropriate positive numbers $t_{in}$, $\gamma_{in}$, $i = 1, \ldots, n$,

$$\exp\{-(c_0 + c_1x + \tfrac{c_2}{x})^\alpha\} = \lim_{n\to\infty}\prod_{i=1}^{n}\Big(1 + \frac{c_0+c_1x+c_2x^{-1}}{t_{in}}\Big)^{-\gamma_{in}}.$$

Now, as $c_0 \geq 2\sqrt{c_1c_2}$,

$$1 + \frac{c_0+c_1x+c_2x^{-1}}{t_{in}} = \frac{c_1}{t_{in}}\cdot\frac{1}{x}\cdot(x+a_{in})(x+b_{in}),$$

where $a_{in}$ and $b_{in}$ are nonnegative. It follows that f is a pointwise limit of functions of the form (5.2.7) and hence, by Theorem 5.2.2, $f \in \mathscr{B}$. □

### 5.3 A Surprise and Its Consequences

The reader may have wondered why the subclass $\mathscr{B}$ of $\mathscr{C}$ was studied so carefully in the previous section. The answer is as simple as it is surprising. We have:

**Theorem 5.3.1.** *The classes $\mathscr{B}$ and $\mathscr{C}$ coincide.*

**Proof.** We know that $\mathscr{B} \subseteq \mathscr{C}$. Let $X$ be an rv such that $X \sim \mathscr{C}$. We shall show that $X \sim \mathscr{B}$. Since $X \sim \mathscr{B}$ iff $1/X \sim \mathscr{B}$, it suffices to show that $1/X \sim \mathscr{B}$. Let $Y_n \sim \text{Gamma}(n,1)$ be independent of X. The pdf of $Y_n/(nX)$ is given by

$$f_n(x) = x^{n-1} \cdot \int e^{-yx} g_n(y)dy \tag{5.3.1}$$

with $g_n(y) = \frac{1}{\Gamma(n)} y^n \cdot f_{nX}(y)$. Obviously $Y_n/(nX) \to 1/X$ in distribution. As $\mathscr{B}$ is closed with respect to weak (non–degenerate) limits, it suffices to verify that $f_n \in \mathscr{B}$ for all $n \geq 1$. If $g_n$ is integrable and thus proportional to a pdf in $\mathscr{C}$ ($\subset \mathscr{T}$ by the Main HCM–theorem), then the integral in (5.3.1) is proportional to the LT of a GGC and hence $f_n \in \mathscr{B}$. If $g_n$ is not integrable, we may consider for $\epsilon > 0$,

$$f_{n,\epsilon}(x) = \tilde{C} \cdot x^{n-1} \int e^{-yx} C \cdot g_n(y) e^{-\epsilon y} dy ,$$

where $C$ is a constant making the integral the LT of a pdf in $\mathscr{C}$ ($\subset \mathscr{T}$) and $1/\tilde{C} = C \cdot \int f_{nX}(y) e^{-\epsilon y} dy$. Now, as is easily verified, $f_{n,\epsilon}$ has the integral 1 and belongs to $\mathscr{B}$. Since $f_{n,\epsilon}(x) \to f_n(x)$ as $\epsilon \to 0$, Theorem 5.2.2 shows that $f_n \in \mathscr{B}$. □

All properties proved true for $\mathscr{C}$ thus hold for $\mathscr{B}$ and vice versa. In particular, $\mathscr{B}$ is closed with respect to multiplication and division of independent rv's; a direct proof does not exist. Further, a pdf of the form $C \cdot f^p$, $p > 0$, belongs to $\mathscr{C}$ if f does. This property of $\mathscr{C}$ is not obvious at all. It easily follows that if a function f is HCM, then so is $f^p$; note that $C \cdot x^{-c} f(x) e^{-\epsilon x}$ is a pdf for some $c \in \mathbb{R}$ and all $\epsilon > 0$. Moreover, the pdf's in $\mathscr{B}$ are analytic in $\mathbb{C} \backslash (-\infty, 0]$ and hence so are the pdf's in $\mathscr{C}$ and the HCM–functions. In the sequel no difference is made between the classes $\mathscr{C}$ and $\mathscr{B}$, and the latter notation is used to be in accord with the notation in earlier works.

There is a nice, very informative characterization of the class $\mathscr{B}$. It can be formulated in different ways.

**Theorem 5.3.2.** *The class $\mathscr{B}$ is the largest class of pdf's $f$ in $\mathscr{T}$ that is closed with respect to multiplication by Gamma densities (and proper normalization).*

**Proof.** Obviously $\mathcal{B}$ is closed with respect to multiplication by Gamma densities. Assume now that $C \cdot y^n e^{-\epsilon y} f_X(y) \in \mathcal{T}$ for all $\epsilon > 0$ and $n \in \mathbb{N}_0$. By rescaling X, we see that also $y^n e^{-\epsilon y} f_{nX}(y)$ is proportional to a pdf in $\mathcal{T}$ for all $\epsilon > 0$ and $n \in \mathbb{N}_0$. Inspection of the proof of Theorem 5.3.1 now shows that $f_X \in \mathcal{B}$. □

Theorem 5.3.2 essentially states that the class $\mathcal{B}$ is the largest subclass of (non–degenerate) distributions in $\mathcal{T}$ that is closed with respect to sampling with probabilities proportional to size x. We may also formulate Theorem 5.3.2 in another way. Let $\mathcal{T}^n$, $n \in \mathbb{N}_0$, be the class of pdf's on $(0,\infty)$ such that, for all $\epsilon > 0$, $C \cdot x^n e^{-\epsilon x} f(x) \in \mathcal{T}$. Obviously $\mathcal{T}^0$ essentially equals $\mathcal{T}$ and, in view of Theorem 4.4.1, $\mathcal{T}^0 \supset \mathcal{T}^1 \supset \mathcal{T}^2 \supset \ldots$ Now Theorem 5.3.2 states that $\mathcal{B} = \mathcal{T}^\infty = \lim_{n\to\infty} \mathcal{T}^n$.

Another reformulation of Theorem 5.3.2 is now given; cf. Theorem 3.1.3.

**Theorem 5.3.3.** *The class $\mathcal{B}$ are those non–degenerate distributions on $(0,\infty)$ for which the mgf $\varphi$ is analytic in $\mathbb{C}\backslash[0,\infty)$ and satisfies*

$$(5.3.2) \qquad \Im m[\varphi^{(n+1)}(s) \cdot \overline{\varphi^{(n)}(s)}] \geq 0 \quad \text{for} \quad \Im m\, s > 0, \; n \in \mathbb{N}_0.$$

The condition (5.3.2) also holds for all degenerate distributions ($\varphi(s) = e^{as}$, $a \geq 0$); these distributions appear as weak limits of distributions in $\mathcal{B}$. When $f \in \mathcal{B}$ and $\varphi$ is smooth up to and on the cut along the positive real axis, $\arg[\varphi^{(n)}(s)]$, $s > 0$, is an increasing function. For $n = 1$, this means that the directed path $\mathbb{R} \ni s \to \varphi(s)$ is *turning to the left*; cf. Section 4.4.

**Remark.** It does not seem possible to characterize the distributions in the class $\mathcal{B}$ in terms of conditions on the U–measure. □

A final characterization describes the K–measures for an $f \in \mathcal{B}$:

**Theorem 5.3.4.** *A pdf $f$ on $(0,\infty)$ belongs to $\mathcal{B}$ iff, for every $u > 0$, $f(uv)f(u/v)$, as a function of $w = v + v^{-1}$ (or $uw$), is proportional to the LT of a GGC.*

**Proof.** Let $f(uv)f(u/v) = \int \exp\{-\lambda uw\}\, K(d\lambda;u)$, where $K(d\lambda;u)$ is proportional to a GGC. Since f is HCM, $f \in \mathcal{B}$. Noting that $w = 0$ for $v = i = \sqrt{-1}$, we also see that $\int K(d\lambda;u) = |f(iu)|^2$. Turning to the necessity, we use the canonical representation for a pdf in $\mathcal{B}$ and find that $f(uv)f(u/v)$ is proportional to

$$\exp\left\{-(b_1u+\frac{b_2}{u})w + \int_{(1,\infty)} \log\left[\frac{(1+y)^2/(uy)}{\frac{y}{u}+\frac{u}{y} + w}\right]\Gamma_1(dy) + \int_{[1,\infty)} \log\left[\frac{(1+y)^2 u/y}{yu+\frac{1}{yu}+ w}\right]\Gamma_2(dy)\right\}$$

Substituting $y' = \frac{y}{u} + \frac{u}{y}$ in the first integral and $y'' = yu + \frac{1}{yu}$ in the second, we see that $f(uv)f(u/v)$, as a function of $w$, is proportional to the LT of a GGC with $a = b_1u + b_2/u$ and with its U–mass on $[u+u^{-1}, \infty)$. □

We now summarize and add a few comments. A pdf f belongs to $\mathscr{B}$ iff one of the following five conditions hold:

*i)* *f is HCM.*

*ii)* *For every $u > 0$, $f(uv)f(u/v)$ is as a function of $w = v + v^{-1}$ proportional to the LT of a GGC.*

*iii)* *f can be represented by (5.2.3).*

*iv)* *$f(x)$ is a limit of functions of the form $C\cdot x^{\beta-1}h(x)$, where h is the LT of a GGC.*

*v)* *$C\cdot x^n e^{-\epsilon x} f(x) \in \mathscr{T}$ for all $\epsilon > 0$ and $n \in \mathbb{N}_0$.*

We have seen that the class $\mathscr{B}$ has many nice structural properties. Densities can be multiplied and be raised to positive powers and the argument can be changed to $x^\alpha$, $|\alpha| \le 1$. The class is closed with respect to multiplication and division of independent rv's and if $X \sim \mathscr{B}$, then $X^q \sim \mathscr{B}$ for $|q| \ge 1$. However, as will be seen in Sections 5.6 and 5.7, the pdf's in $\mathscr{B}$ are not infinitely divisible with respect to multiplication of rv's and $\mathscr{B}$ is not closed with respect to convolution roots although $\mathscr{B} \subset \mathscr{T}$.

A *probability mechanism* leading to the *full* class $\mathscr{B}$ is still lacking though likely to be somewhere. Finding such a model is an important problem.

## 5.4 Decreasing HCM–functions

This section contains a result that will be very useful in Chapter 6.

Let f be a function on $(0,\infty)$ which is HCM. As seen in Section 5.1, the function $x^{-c}f(x)$ is then bounded for some $c \in \mathbb{R}$. Thus, for any $\epsilon > 0$, $C\cdot x^{-c}e^{-\epsilon x}f(x)$ is not only HCM but also a pdf in $\mathscr{B}$. Dividing by $x^{-c}$ and letting $\epsilon \to 0$, we conclude from Theorem 5.2.2 that, as expected, f also admits the canonical representation

$$f(x) = C\cdot x^{\beta-1}\cdot\exp\left\{-b_1x - \frac{b_2}{x} + \int_{(1,\infty)} \log\left(\frac{y+1}{y+x}\right)\Gamma_1(dy) + \int_{[1,\infty)} \log\left(\frac{y+1}{y+x^{-1}}\right)\Gamma_2(dy)\right\}.$$

Of particular interest is the case when f is decreasing.

**Theorem 5.4.1.** *A function f is HCM and $f(0+) > 0$ iff f is the LT of a widened GGC (and thus decreasing). Thus f is HCM and $f(0+) = 1$ iff f is the LT of a GGC.*

**Proof.** We know that the "if" parts hold. Now let $f$ be HCM and $f(0+) > 0$. By the logconcavity of $f(e^y)$, $f$ is monotone in a neighbourhood of zero. Since then

$$\liminf_{t\to 0} f(tx)/f(t) \geq 1, \quad 0 < x \leq 1, \tag{5.4.1}$$

and, by the canonical representation,

$$f(tx)/f(t) \sim x^{\beta-1}\cdot\exp\{-b_2\frac{x^{-1}-1}{t} + \log x \cdot \Gamma_2([1,\infty))\} \qquad \text{as } t \to 0,$$

it follows that $b_2 = 0$ and that $c_0 = -(\beta-1+\Gamma_2([1,\infty)) \geq 0$. Now

$$x^{\beta-1}\cdot\exp\{\int_{[1,\infty)} \log(\tfrac{y+1}{y+x^{-1}})\Gamma_2(dy)\} = x^{-c_0}\cdot\exp\{\int_{[1,\infty)} \log(\tfrac{y+1}{y+x^{-1}}\cdot\tfrac{1}{x})\Gamma_2(dy)\}.$$

The substitution $y' = y^{-1}$ in the last expression shows that $f$ can be represented as

$$f(x) = C\cdot x^{-c_0}\cdot\exp\{-b_1x + \int \log(\tfrac{y+1}{y+x})\Gamma(dy)\}. \tag{5.4.2}$$

Here the measure $\Gamma$ equals $\Gamma_1$ on $(1,\infty)$ and on $(0,1]$ it equals the measure induced by $\Gamma_2$ and the transformation above. We see that $f$ is the LT of a *widened GGC*.

If $f(0+) = 1$, then it is easily seen from (5.4.2) that $c_0 = 0$ and $C\cdot\exp\{\int \log(\frac{y+1}{y})\Gamma(dy)\} = 1$. Hence $f$ is the LT of a GGC. □

**Corollary 1.** *A function $f$ is HCM and has limit 1 as $x \to \infty$, iff $g(x) = f(x^{-1})$ is the LT of a GGC.*

**Corollary 2.** *If $f$ is decreasing and HCM, then so is $f(x+\delta)$ for any $\delta \geq 0$.*

Corollary 2 is most obvious from the fact that a decreasing HCM–function $f(x)$ can be obtained as a limit of products of functions of the type $C\cdot(y+x)^{-\gamma}$ $(y \geq 0, \gamma > 0)$. □

## 5.5 Creation of New Densities in $\mathcal{B}$ by Composition

Let $\check{\mathcal{T}}$ denote the class of LT's of GGC's. A pdf of the form $f(x) = C\cdot x^{\beta-1}h(x)$, $x > 0$, where $\beta > 0$ and $h \in \check{\mathcal{T}}$, belongs to $\mathcal{B}$ by definition. Now, if $h \in \check{\mathcal{T}}$, then $h(\xi(x))$ belongs also to $\check{\mathcal{T}}$ for a rather large class of functions $\xi$ including $\xi(x) = x^\alpha$, $0 < \alpha \leq 1$, and, more generally,

$$\xi(x) = \sum_{k=1}^{n} c_k x^{\alpha_k}, \qquad c_k > 0,\ 0 < \alpha_k \leq 1. \tag{5.5.1}$$

This was found in Section 3.3 although expressed for mgf's. It follows that also a pdf of the form $f(x) = C \cdot x^{\beta-1}h(\xi(x))$ belongs to $\mathscr{B}$. There is a simple way of finding an appropriate class of functions $\xi$ as will now be seen.

**Theorem 5.5.1.** *Let $\xi(x) > 0$, $x > 0$, be a function satisfying $\xi(0+) = 0$ and such that*

*i) For each $u > 0$, $g_1(w) = \xi(uv) + \xi(u/v)$ has a CM derivative with respect to $w = v + v^{-1}$.*

*ii) For each $u > 0$, $g_2(w) = \xi(uv)\xi(u/v)$ has a CM derivative with respect to $w$.*

*Then every pdf of the form $f(x) = C \cdot x^{\beta-1}h(\xi(x))$, $\beta > 0$, $h \in \check{\mathscr{T}}$, belongs to $\mathscr{B}$.*

**Proof.** It suffices to consider the case when $h(x) = (1+cx)^{-\gamma}$, $c > 0$, $\gamma > 0$. Then

$$f(uv)f(u/v) = C^2u^{2(\beta-1)} \cdot (1 + cg_1(w) + c^2g_2(w))^{-\gamma}.$$

Since $(1+x)^{-\gamma}$ is CM and $cg_1(w) + c^2g_2(w)$ has a CM derivative, it follows that $f$ is HCM. Thus $f \in \mathscr{B}$. □

**Example 5.5.1.** Let $\xi(x)$ be of the form (5.5.1). Now the derivative of

$$g_1(w) = \xi(uv) + \xi(u/v) = \Sigma\, c_k u^{\alpha_k}(v^{\alpha_k} + v^{-\alpha_k})$$

is certainly CM since the derivative of $g(w) = v^{\alpha}+v^{-\alpha}$ is. Further, for the same reason,

$$g_2(w) = \xi(uv) \cdot \xi(u/v) = \Sigma\, c_k^2\, u^{2\alpha_k} + \sum\sum_{j<k} c_j c_k u^{\alpha_k+\alpha_j}(v^{\alpha_k-\alpha_j} + v^{\alpha_j-\alpha_k})$$

has a CM derivative. Hence (i) and (ii) hold. In particular, it follows that pdf's of the forms $f(x) = C \cdot x^{\beta-1}(1 + \Sigma c_k x^{\alpha_k})^{-\gamma}$ and $f(x) = C \cdot x^{\beta-1}\exp\{-\Sigma c_k x^{\alpha_k}\}$ are in $\mathscr{B}$. The last pdf belongs to $\mathscr{B}$ even if some of the $\alpha_k$'s are negative, provided that $|\alpha_k| \le 1$. □

Theorem 5.5.1 can be partially improved. We have

**Theorem 5.5.2** ***(Composition theorem).*** *Let $\xi(x)$, $x > 0$ be* ***increasing*** *and such that $\xi(0+) = 0$. Then $h(\xi(x))$ is the LT of a GGC whenever $h$ is* ***iff***

$$(i')\ e^{-\xi(x)} \text{ is HCM} \quad \text{and} \quad (iii')\ 1/\xi(x) \text{ is HCM},$$

*or, equivalently, cf. Theorem 5.5.1, iff*

$$(i)\ g_1'(w) \text{ is CM} \quad \text{and} \quad (iii)\ \frac{d}{dw}\log g_2(w) \text{ is CM}.$$

**Proof.** To show that for instance (iii') ⇒ (iii), we recall that $1/\xi(x)$ is HCM iff $1/(\xi(x))^p$ is HCM for all $p > 0$. Hence (iii') implies that $\exp\{-p\cdot\log(\xi(uv)\xi(u/v))\} = \exp\{-p\cdot\log(g_2(w))\}$ is CM. Since this is true for all $p > 0$, (iii) easily follows. Analogously, (i') ⇒ (i). It is evident that (i) ⇒ (i') and (iii) ⇒ (iii').

As to the first part of the theorem, the "only if" part is rather obvious. To prove the "if" part, we have to use Theorem 3.3.1. Let $\zeta(s) = -\xi(-s)$ and $\varphi(s) = h(-s)$. Since $1/\xi$ is decreasing and HCM, $1/\xi$ is the LT of a widened GGC; cf. Section 5.4. Thus $-1/\zeta(s)$ is the mgf of a widened GGC. It follows that $\Im m[\zeta'(s)\overline{\zeta(s)}] \le 0$ for $\Im m\, s > 0$. By Theorem 3.3.1, $\varphi(\zeta(s))$ is the mgf of a GGC. Equivalently, $h(\xi(x))$ is the LT of a GGC. □

### 5.6 Multiplication of $\mathscr{B}$–variables

There are many consequences of the fact that $\mathscr{B}$ is closed with respect to multiplication of independent rv's. Some are revealed here. Others appear in Chapters 6–8.

**Example 5.6.1** ***(Lognormal distribution revisited).*** Let $X_{jn}$, $j = 1, \ldots, n$, be iid rv's such that $X_{jn} \sim \mathrm{Gamma}(n/\tau, n/\tau)$, $\tau > 0$. Then $X_n = e^{\tau/2}\cdot\Pi_{j=1}^n X_{jn}$ has asymptotically a Lognormal distribution with parameters $\mu = 0$ and $\sigma^2 = \tau$. This is seen by taking logarithms and using an appropriate version of the central limit theorem for triangular arrays. For $Y_{jn} = \log X_{jn}$ we have, as $n \to \infty$,

$$\mu_{jn} = E[Y_{jn}] \sim E[X_{jn} - 1 - (X_{jn}-1)^2/2] = -\tfrac{1}{2}\,\mathrm{Var}[X_{jn}] = -\tfrac{1}{2}\cdot\tfrac{\tau}{n}$$

and $\mathrm{Var}[Y_{jn}] \sim \mathrm{Var}[X_{jn}-1] = \tau/n$. Further, for any $c > 0$, by tedious calculations, $E[|Y_{jn}-\mu_{jn}|^{2+c}] \sim \mathrm{constant}\cdot n^{-d}$, where $d = 1 + c/2$. Thus, the well–known Ljapounov condition is satisfied. Again it is found that the Lognormal distribution is in $\mathscr{B}$. □

**Remark** ***(LogCauchy distribution).*** Let $f_Y(y) = \frac{1}{\pi}\cdot(1+y^2)^{-1}$, $y \in \mathbb{R}$, and consider $X = e^Y$ with pdf $f_X(x) = \frac{1}{\pi}\cdot\frac{1}{x}\cdot(1+(\log x)^2)^{-1}$, $x > 0$. As the Cauchy distribution is a limit distribution for sums of iid rv's, one may guess that $f_X \in \mathscr{B}$ also. However, it does not even belong to $\mathscr{T}$. In fact, if it did then its total U–measure would be zero by Theorem 4.1.4. The same argument shows that there is in $\mathscr{T}$ no LogStable distribution induced by a non–Normal symmetric Stable distribution. □

**Example 5.6.2** ***(Stable distribution).*** Let $X_\alpha$ be a positive Stable rv with LT $\phi(s) = \exp\{-s^\alpha\}$. We shall show that $X_\alpha \sim \mathscr{B}$ at least for $\alpha = n^{-1}$, $n = 2, 3, \ldots$

Let $Y \sim \text{Gamma}(1,1)$ be independent of $X_\alpha$. Then

$$\Pr(\{(Y/X_\alpha)^\alpha > x\}) = \Pr(\{Y > x^{1/\alpha}X_\alpha\}) = \mathbf{E}[\exp\{-x^{1/\alpha}X_\alpha\}] = e^{-x},$$

i.e., $(Y/X_\alpha)^\alpha \stackrel{d}{=} Y$ or $X_\alpha/Y \stackrel{d}{=} Y^{-1/\alpha}$. It follows that

$$\mathbf{E}[X_\alpha^{-k}] = \mathbf{E}[Y^{k/\alpha}]/\mathbf{E}[Y^k] = \Gamma(\tfrac{k}{\alpha}+1)/\Gamma(k+1), \quad k > 0, k \in \mathbb{R}.$$

Let now $Y_1, Y_2, \ldots, Y_{n-1}$ be independent Gamma variables such that $Y_j \sim \text{Gamma}(\frac{j}{n}, 1)$ and consider $Z_n = n^n Y_1 Y_2 \cdot \ldots \cdot Y_{n-1} \sim \mathscr{B}$. Then, for $k \in \mathbb{N}_0$,

$$\mathbf{E}[Z_n^k] = n^{nk} \cdot \frac{\Gamma(k+\frac{1}{n})\Gamma(k+\frac{2}{n})\cdot \ldots \cdot \Gamma(k+\frac{n-1}{n})}{\Gamma(\frac{1}{n})\Gamma(\frac{2}{n})\cdot \ldots \cdot \Gamma(\frac{n-1}{n})} = n^{nk} \prod_{i=1}^{n-1} \prod_{j=1}^{k} (\tfrac{i}{n} + j-1) = \frac{\Gamma(nk+1)}{\Gamma(k+1)},$$

where the last expression is obtained by a simple rearrangement of the factors. By slightly more complicated rearrangements one can show that $\mathbf{E}[Z_n^k] = \Gamma(nk+1)/\Gamma(k+1)$ also for $k = \ell/n$, $\ell \in \mathbb{N}_0$. See Formula 6.1.20 in Abramowitz & Stegun (1964). Thus, by some reflection, the moments of $Z_n^{n^{-1}}$ and $(X_\alpha^{-1})^{n^{-1}}$, $\alpha = n^{-1}$, coincide. By Carleman's condition (Feller, 1971, pp. 227–228) these moments determine the distribution uniquely and it follows that $X_{n^{-1}} \stackrel{d}{=} 1/Z_n \sim \mathscr{B}$. Much indicates that $X_\alpha \sim \mathscr{B}$ for all $\alpha \leq 1/2$. Unfortunately no complete proof is available of that; cf. Sections 5.7, 5.8, and 7.2. □

**Example 5.6.3** ***(Products of powers of Gamma variables).*** Let $\mathscr{G}$ denote the class of distributions of rv's of the form $X = Y_1^{q_1} \cdot Y_2^{q_2} \cdot \ldots \cdot Y_n^{q_n}$, where $|q_j| \geq 1$ and the $Y_j$'s are independent Gamma variables. Obviously $\mathscr{G} \subseteq \mathscr{B}$ and $\mathscr{G}$ is closed with respect to multiplication and division of independent rv's. Since, as will be seen in Example 7.2.3, the logarithm of a Gamma variable is infinitely divisible (ID), the distributions in $\mathscr{G}$ are multiplicatively ID. Even if weak (non–degenerate) limits are included in $\mathscr{G}$, it is a strict subclass of $\mathscr{B}$ since not all distributions in $\mathscr{B}$ are multiplicatively ID. In fact, if $X \sim C \cdot \exp\{-x - x^{-1}\} \in \mathscr{B}$, then the pdf of $\log X$ equals $C \cdot \exp\{x - e^x - e^{-x}\}$ and it tends more rapidly to zero as $|x| \to \infty$ than is possible for a non–Normal ID pdf; cf. Lukacs (1983, pp. 69–70). □

### 5.7 $\mathscr{B}$–distributions with Explicit Laplace Transforms

Although by Theorem 5.3.3 the $\mathscr{B}$–distributions can be characterized in terms of conditions on the mgf or the Laplace transform $\phi$, which is preferred in this section, there is no canonical representation formula for the LT of a $\mathscr{B}$–distribution. However, there are $\mathscr{B}$–distributions for which the LT has a simple expression but the pdf is less explicit. In this section such distributions are studied. The results will be used in Chapters 7 and 8.

Let $\breve{\mathscr{B}}$ denote the class of LT's of distributions in $\mathscr{B}$. The following simple operations map $\breve{\mathscr{B}}$ into $\breve{\mathscr{B}}$:

i) *(Differentiation and Translation)* $\phi \to \phi^{(n)}(s+\delta)/\phi^{(n)}(\delta)$, $n \in \mathbb{N}_0$, $\delta > 0$.

ii) *(Integration)* $\phi \to C \cdot e^{\delta s} \cdot \int_s^\infty (t-s)^{\gamma-1} e^{-\delta t} \phi(t)dt$, $\gamma, \delta > 0$.

iii) *(Composition)* $\phi \to \phi(s^\alpha)$, $\alpha = 1/k$, $k = 1, 2, 3, \ldots$

Here (i) and (ii) correspond to multiplication of the pdf by $x^n e^{-\delta x}$ and $(x+\delta)^{-\gamma}$, respectively. If $Y$ and $X$ are independent rv's with LT's $\phi$ and $\exp\{-s^\alpha\}$, respectively, then the LT corresponding to $Y^{1/\alpha}X$ equals $\phi(s^\alpha)$. We know from Section 5.6 that $X \sim \mathscr{B}$ for $\alpha^{-1} = 2, 3, \ldots$ and therefore $Y^{1/\alpha}X \sim \mathscr{B}$ provided that $Y \sim \mathscr{B}$; hence (iii) follows. Again it would be desirable that $X \sim \mathscr{B}$ for all $\alpha \leq 1/2$.

In particular $\phi(s) = (1 + s^{1/k})^{-\gamma}$, $k \in \mathbb{N}$, belongs to $\breve{\mathscr{B}}$ but there is no explicit expression for the pdf for $k \neq 1$. The LT $\phi(s) = \exp\{-\sqrt{s+\delta} + \sqrt{\delta}\}$ corresponds to the inverse Gaussian distribution with pdf $f(x) = C \cdot x^{-3/2} \exp\{-\delta x - \frac{1}{4} \cdot x^{-1}\}$. Since, for $k = 2, 3, \ldots$, $\phi(s) = \exp\{-(s+\delta)^{1/k} + \delta^{1/k}\}$ belongs to $\breve{\mathscr{B}}$ so does, by (i),

$$\phi_1(s) = \phi'(s)/\phi'(0) = (1 + \tfrac{s}{\delta})^{\frac{1}{k}-1} \phi(s) .$$

Obviously the corresponding distribution has a Gamma$(1-k^{-1}, \delta)$–distribution as convolution factor, i.e., the U–measure has an atom of size $1-k^{-1}$ at $t = \delta$.

**Remark.** One can verify by explicit calculation that for $k = 2$ the mgf $\psi_m(s) = (\phi_1(-s))^{1/m}$, $m \geq 1$, has the property that $\Im m[\psi_m''(s)\overline{\psi_m'(s)}]$ is negative for some $s > 0$ (on the upper side of the cut) provided that $m$ is large enough. So, in view of Theorem 5.3.3, $\mathscr{B}$ *is not closed with respect to convolution roots.* □

Now some less trivial members of $\breve{\mathscr{B}}$ are considered.

**Theorem 5.7.1.** *For* $0 < \epsilon < 1$ *and* $\delta \geq 0$, $\phi_\epsilon(s) = \frac{\delta-1}{\delta^\epsilon-1} \cdot \frac{(s+\delta)^\epsilon-1}{s+\delta-1} \in \breve{\mathscr{B}}$. *Moreover,* $\phi_\epsilon(s) \in \breve{\mathscr{B}}$ *for* $-1 < \epsilon < 0$ *and* $\delta > 0$.

**Proof.** By tedious calculations it can be verified that $\phi_\epsilon(s) = \int e^{-sx} f(x)dx$ , where

$$f(x) = C\cdot e^{-\delta x} \int_0^\infty \frac{1}{y}\cdot(1+\frac{x}{y})^{-1-\epsilon} y^{-\epsilon} e^{-y}\, dy .$$

For $\epsilon > 0$ and $\delta = 0$ we see that f is the pdf of the rv $Z = Y\cdot X_1/X_2$, where $Y \sim \text{Gamma}(1-\epsilon, 1)$, $X_1 \sim \text{Gamma}(1, 1)$, and $X_2 \sim \text{Gamma}(\epsilon, 1)$ are independent rv's. Since $Z \sim \mathcal{B}$, the theorem holds for $\epsilon > 0$ and all $\delta \geq 0$. Let now $\epsilon < 0$. The integral is of the form $\int g(x/y)h(y)dy$, where g and h both are HCM. Hence f is HCM by essentially property (vii) in Section 5.1. Thus $f \in \mathcal{B}$. □

**Remark.** For $\epsilon > 0$, $\phi_{-\epsilon}(s) = (1+\frac{s}{\delta})^{-\epsilon}\phi_\epsilon(s)$; again we have an example of a distribution in $\mathcal{B}$ with an atom for its U–measure. □

**Corollary 1.** *The function* $\phi(s) = \frac{\log(1+s)}{s}$ *belongs to* $\check{\mathcal{B}}$.

**Proof.** The corollary follows from the fact that $\frac{\log(1+s)}{s}$ is a limit of $\frac{1}{\epsilon}\frac{(s+1)^\epsilon-1}{s+1\,-\,1}$ as $\epsilon \downarrow 0$. The corresponding pdf is $f(x) = e^{-x} \int e^{-xy}(1+y)^{-1}dy$ . □

**Corollary 2.** *Let* $X \sim \text{Gamma}(\beta, \lambda)$, $\beta \geq 1$. *Then* $(1+X)^q-1 \sim \mathcal{B}$, $q \geq 1$, *and* $e^X-1 \sim \mathcal{B}$.

**Proof.** Let $Y = (1+X)^q-1$ and set $\epsilon = q^{-1} \leq 1$. Then

$$f_Y(x) = \epsilon\cdot(1+x)^{\epsilon-1} f_X((1+x)^\epsilon-1) = C\cdot x^{\beta-1}h(x),$$

where

$$h(x) = \left(\frac{(1+x)^\epsilon-1}{x}\right)^{\beta-1}(1+x)^{\epsilon-1}\exp\{-\lambda((1+x)^\epsilon-1)\}.$$

Mainly by Theorem 5.7.1, h is the LT of a GGC for $\beta \geq 1$ and hence $Y \in \mathcal{B}$. It follows also that $(1+\frac{X}{q})^q - 1 \sim \mathcal{B}$. Letting $q \to \infty$, we get $e^X-1 \sim \mathcal{B}$. □

Corollary 2 has some importance as will be seen in the next chapter. A final example may illustrate its usefulness.

**Example 5.7.1** ***(Lognormal distribution revisited).*** Let $X \sim \text{Gamma}(\beta, 1)$, $\beta \geq 1$. By Corollary 2, $\exp\{X/\sqrt{\beta}\} - 1 \sim \mathcal{B}$ . By multiplying by $\exp\{-\sqrt{\beta}\}$, we find that $\exp\{(X-\beta)/\sqrt{\beta}\} - \exp\{-\sqrt{\beta}\} \sim \mathcal{B}$. Letting then $\beta \to \infty$, we see again that the (standard) Lognormal density belongs to $\mathcal{B}$. □

### 5.8 Complementary Results and Notes

I (***Strongly unimodal*** $\mathscr{B}$***–distributions***). As seen, the pdf's in $\mathscr{B}$ are unimodal. However, they are not always strongly unimodal, i.e., logconcave. It is not hard to show that a pdf in $\mathscr{B}$ is logconcave iff, in terms of the canonical representation, $b_1 > 0$ and $\beta - 1 \geq \Gamma_1((1,\infty))$.

II ($\mathscr{B}$***–distributions with Gamma components***). The class $\mathscr{B}$ is not closed under convolution. However, we have seen three examples of $\mathscr{B}$–distributions with a Gamma distribution as convolution factor. These distributions are somewhat specific. Recalling that for a distribution in $\mathscr{B}$ the path $\mathbb{R} \ni s \to \varphi(s)$ (or $\varphi(s+i\cdot\delta)$, $\delta > 0$) is turning to the left, we see by some geometry that the U–measure can have at most one atom. It must be located at the left–extremity of the U–measure. There is a general method of constructing examples of pdf's in $\mathscr{B}$ of the specific kind:

Let $f_1$ be a Gamma($\beta,\lambda$)–density. Then, as can be shown by the help of (xi) in Section 5.1 and some transformations ($x \to x^{-1}$ and $y \to y^{-1}$), the convolution $f = f_1 \star f_2$ is HCM (i.e., in $\mathscr{B}$) if $x^{\beta+1}e^{\lambda x}f_2(x)$ is *HCM and increasing.* A simple example is provided by

$$f(x) = x\cdot e^{-x}/(1+x) = e^{-x} \star e^{-x}/(1+x)^2.$$

III (***Exponentiation of Gamma variables***). Let $f$ be the pdf of $Y = e^X - 1$, where $X \sim \text{Gamma}(\beta,\lambda)$. We have seen that $f \in \mathscr{B}$ for $\beta \geq 1$. However, $f \notin \mathscr{B}$ for $\beta < 1$. In fact, if $f \in \mathscr{B}$, then since $f(x) = C\cdot(\log(1+x))^{\beta-1}(1+x)^{-\lambda-1}$ is decreasing, $f$ must be the LT of a widened GGC, i.e., $\psi(s) = f(-s)$ must satisfy $\Im m[\psi'(s)/\psi(s)] \geq 0$ for $\Im m\, s > 0$. However, as is not hard to verify, $\Im m[\psi'(s)/\psi(s)] < 0$ for $s > 1$ (on the upper side of the cut).

IV (***Stable distribution***). It is conjectured that $\varphi(s) = \exp\{-(-s)^\alpha\}$ is the mgf of a distribution in $\mathscr{B}$ for $\alpha \leq 1/2$. Support for this conjecture is given below. Using the Taylor expansion of $\varphi(s)$ with respect to $(-s)^\alpha$ and differentiating, we get

$$\psi_n(s) = (-s)^{n-\alpha}\varphi^{(n)}(s) = (-1)^n \sum_{j=1}^{\infty} \frac{(-1)^j}{j!} j\alpha\cdot(j\alpha-1)\cdot\ldots\cdot(j\alpha-n+1)\cdot(-s)^{(j-1)\alpha}.$$

It suffices to show that, for each $n \geq 1$, $\psi_n(s)$ is proportional to the mgf of a GGC. By Theorem 3.3.1 (Composition Theorem) it is enough to verify that $\chi_n(s) = \psi_n(-(-s)^{1/(2\alpha)})$ has the same property. It can be shown that this is true if $\Im m[\chi_n'(s)\overline{\chi_n(s)}] > 0$ for $s > 0$; cf. Theorem 3.1.4. By a lot of algebra we get, for $s > 0$,

$$\Im m[\chi_n'(s)\overline{\chi_n(s)}] = \frac{1}{2}\sum_{m=1}^{\infty} P_{nm}(\alpha)\cdot s^{m-1.5},$$

where

$$P_{nm}(\alpha) = \sum_{j+k=2m+1} k\cdot\frac{k\alpha\cdot(k\alpha-1)\cdot\ldots\cdot(k\alpha-n+1)}{k!}\,\frac{j\alpha\cdot(j\alpha-1)\cdot\ldots\cdot(j\alpha-n+1)}{j!}\,\epsilon_{k-j}$$

with $\epsilon_{k-j} = \Im m[i^{k-j}]$; $i = \sqrt{-1}$. The polynomial $P_{nm}$ has degree $2n$ and can be shown to vanish identically for $m > n$. It has a zero of order $2m$ at $\alpha = 0$ and moreover a simple zero at $\alpha = 1/2$ for $m < n$. It seems as though even $P_{nm}(\alpha) > 0$ for $0 < \alpha < 1/2$ and $m \leq n$. By extensive numerical calculations this has been shown to be true at least for $n \leq 150$.

**Notes.** Most of the results in this chapter are the author's. The nice and important Example 5.6.2 (Stable distribution) is due to Williams (1977).

The term "hyperbolic complete monotonicity" has not been used in any other publication. The systematic use of this notion has much simplified the theory. Theorem 5.1.1 (Multiplication Theorem) appeared in B1990. The results in Sections 5.2 and 5.3 are improvements of results in B1979c.

Theorem 5.4.1 appeared in B1990. Theorems 5.5.1 and 5.5.2 (Composition Theorem) are new but have their origin in B1979a,c. Ebbesen (1983) pointed out an error in the proof in B1979c and corrected it; cf. B1984. Products of powers of independent Gamma variables were considered in B1979b. Earlier, Goovaerts, D'Hooge & de Pril (1977b) proved that the distribution of a product of two independent Gamma variables is a GGC. The results in Section 5.7 are partly new; some appeared in B1981.

# 6. BACK TO $\mathcal{T}$

The results in Chapter 5 have many spin–off effects that concern the full class $\mathcal{T}$. These are the focus of this chapter. In Section 6.1 real characterizations of some transforms are given in terms of hyperbolic complete monotonicity (HCM) and then, in Section 6.2, the fruit of this work is harvested. It is shown that $\mathcal{T}$ also has several multiplicative properties. Moreover, the effect of certain transformations of rv's with distributions in $\mathcal{T}$ is studied. In Section 6.3 essentially some large subclasses of $\mathcal{T}$ are considered. A short summary of Chapters 5 and 6 is given in Section 6.4, and an attempt is made to widen the perspective.

## 6.1 Real Characterizations of Some Transforms

We know from Section 5.4 that the LT of a GGC can be characterized as follows.

**Theorem 6.1.1** ***(Real characterization).*** *A function $\phi(s)$ on $(0,\infty)$ is the LT of a GGC iff $\phi(0+) = 1$ and $\phi$ is HCM.*

Some particular distributions given by their LT's, as, e.g., the Stable distribution, can be shown to be GGC's by Theorem 6.1.1. It follows also easily that for instance $\phi(s^\alpha)$, $0 < \alpha \leq 1$, is the LT of a GGC whenever $\phi$ is; cf. Section 5.5. However, as we shall see in Section 6.2, there are more far–reaching applications of the theorem. Theorem 6.1.1 makes the difficult complex GGC–theory partly superfluous in the sequel.

**Remark.** Given Theorem 6.1.1 we can close a circle and show that $\mathcal{B} \subset \mathcal{T}$ in a simple way. Let $f \in \mathcal{B}$ and let $\phi(s) = \int e^{-sx}f(x)dx$. An application of property (vii) in Section 5.1 shows that $\phi$ is HCM. Hence $f \in \mathcal{T}$. □

For the class $\mathcal{B}$ there is an analogue of Theorem 6.1.1. It is not as useful as Theorem 6.1.1.

**Theorem 6.1.2.** *Let $\phi$ be the LT of a non–degenerate probability distribution on $(0,\infty)$. The distribution belongs to $\mathcal{B}$ iff, for every $n \in \mathbb{N}_0$, $(-1)^n\phi^{(n)}$ is HCM.*

**Proof.** Let $f \in \mathcal{B}$. Since $(-1)^n\phi^{(n)}(s) = \int e^{-sx}x^n f(x)dx$ and $x^n f(x)$ is HCM, it follows by property (vii) in Section 5.1 that $(-1)^n\phi^{(n)}$ is also HCM. Conversely, if $(-1)^n\phi^{(n)}$ is HCM, then since $(-1)^n\phi^{(n)}$ is decreasing, we have that, for each $\epsilon > 0$, $(-1)^n\phi^{(n)}(s+\epsilon)$ is proportional to the LT of a GGC; cf. Section 5.4. Hence

$C\cdot x^n e^{-\epsilon x} f(x) \in \mathcal{T}$. Since this holds for all $n \in \mathbb{N}_0$, it follows that $f \in \mathcal{B}$; cf. Theorem 5.3.2 and its proof. □

Theorem 6.1.1 can be transformed into a real characterization of the Stieltjes transform of a nonnegative measure, i.e., a function of the form

$$\text{(6.1.1)} \qquad \psi(s) = a + \int_{[0,\infty)} \frac{1}{t+s}\, U(dt), \quad a \geq 0, \; U(dt) \geq 0.$$

(One may think of $a$ as the mass for $U(dt)/t$ at $t = \infty$.) When $\psi(0+) = 1$, this is the LT of a mixture of Exponential distributions. If $\phi$ is the LT of a widened GGC, then $\psi(s) = -\phi'(s)/\phi(s)$ has the form (6.1.1).

**Theorem 6.1.3.** *A (locally integrable) function $\psi$ on $(0,\infty)$ has the form (6.1.1) iff $\liminf_{s\downarrow 0} s\psi(s) \geq 0$ and, for each $u > 0$, $g_u(w) = (v\psi(uv) - v^{-1}\psi(u/v))/(v - v^{-1})$ is a CM function of $w = v + v^{-1}$.*

**Proof.** The "only if" part is left to the reader to verify. Turning to the "if" part, we set $\phi(x) = \exp\{-\int_1^x \psi(s)ds\}$. As the derivative of $\int_1^{uv}\psi(s)ds + \int_1^{u/v}\psi(s)ds$ with respect to $w$ equals $g_u(w)$, it follows that $\phi(uv)\phi(u/v)$ is CM in $w$, i.e., $\phi$ is HCM. For $x < 1$,

$$\liminf_{t\to 0} \frac{\phi(tx)}{\phi(t)} = \exp\{\liminf_{t\to 0} \int_{tx}^{t} \psi(s)ds\} \geq e^0 = 1,$$

since $\liminf_{s\to 0} s\psi(s) \geq 0$. The results in Section 5.4 show that $\phi$ $(= f)$ has the form (5.4.2). Since $\psi(s) = -\frac{d}{ds}\log\phi(s)$, the desired result then follows by just a change of notation: $y \to t$, $\Gamma(dy) \to U(dt)$, $c_0 \to U(\{0\})$, and $b_1 \to a$. □

**Example 6.1.1** *(A new GGC).* We shall show that $\phi(s) = (1 + \sqrt{s}/n))^n \cdot \exp\{-\sqrt{s}\}$, $n > 0$, is the LT of a GGC by applying Theorem 6.1.3. We have to prove that $-\phi'(s)/\phi(s) = (2n)^{-1}/(1 + \sqrt{s}/n)$ is the Stieltjes transform of a nonnegative measure. It suffices to show that $\psi(s) = 1/(1 + \sqrt{s})$ is such a transform. Recalling the definition of $g_u(w)$ and making some calculation, we get

$$g_u(w) = \left(1 + \frac{\sqrt{u}}{v^{1/2}+v^{-1/2}}\right) \cdot \frac{1}{1 + u + \sqrt{u}\cdot(v^{1/2}+v^{-1/2})}.$$

Since $v^\alpha + v^{-\alpha}$ $(\alpha \leq 1)$ has a CM derivative with respect to $w$, $g_u(w)$ is CM. Thus Theorem 6.1.3 gives the desired result. □

## 6.2 Multiplicative Properties of $\mathcal{T}$

In this section we apply Theorem 6.1.1 to prove some important and deep properties of the GGC's. The results have nice and surprising consequences.

The class $\mathcal{B}$ is closed with respect to multiplication of independent rv's. There is a dual result for $\mathcal{T}$.

**Theorem 6.2.1** (***Multiplication theorem for*** $\mathcal{T}$). *Let* $X \sim \mathcal{T}$ *and* $Y \sim \mathcal{B}$ *be independent rv's. Then* $X \cdot Y \sim \mathcal{T}$ *and* $X/Y \sim \mathcal{T}$.

**Proof.** Since $Y \sim \mathcal{B}$ iff $1/Y \sim \mathcal{B}$, it suffices to show that $X \cdot Y \sim \mathcal{T}$. Its LT is given by $\phi_{XY}(s) = \int \phi_X(sy) f_Y(y) dy$. Since $\phi_X$ and $f_Y$ both are HCM, it follows by property (vii) in Section 5.1 that so is $\phi_{XY}$. Hence, by Theorem 6.1.1, $X \cdot Y \sim \mathcal{T}$. □

**Remark.** It is not true that in general $Y/X \sim \mathcal{T}$ for independent rv's $Y \sim \mathcal{B}$ and $X \sim \mathcal{T}$. In fact, if it were, then $1/X \sim \mathcal{T}$ also. However, that is not true if $X$ has left–extremity $> 0$ as then $1/X$ is bounded. □

In particular we see that a scale mixture of Gamma($\beta,\cdot$) distributions ($\beta$ fixed) with a GGC as scale mixing distribution is a GGC. Infinite divisibility (ID) of general scale mixtures of Gamma($\beta,\cdot$) distributions was studied by Steutel (1970). His conjecture that such mixtures are ID when $\beta = 2$ (and hence when $\beta \leq 2$ also) has remained unproved for a long time. Recently, a very complicated proof of it has been presented by Kristiansen (1991).

The most important consequence of Theorem 6.1.1 is obtained by considering $X/Y_n$ with $Y_n \sim$ Gamma(n, n), n = 1, 2, ... The pdf of $X/Y_n$ $(= Y_n^{-1} \cdot X)$ equals

$$f_n(x) = \frac{1}{\Gamma(n+1)} \left(\frac{n}{x}\right)^{n+1} \int \exp\{-\tfrac{n}{x} y\}\, y^n F_X(dy) = \frac{(-1)^n}{n!} \left(\frac{n}{x}\right)^{n+1} \phi_X^{(n)}\left(\frac{n}{x}\right). \tag{6.2.1}$$

Every pdf of this form (with $X \sim \mathcal{T}$) is thus a GGC. Now $f_n$ tends weakly to the distribution of $X$ as $n \to \infty$ and hence, by, e.g., Theorem 4.1.5, also pointwise to the pdf of $X$ when $X$ is non–constant. This convergence is known as *the real inversion formula for the LT*. Thus, we have a subclass of explicit pdf's in $\mathcal{T}$ such that the subclass is *dense* in $\mathcal{T}$. Sometimes a given pdf can be verified to be a GGC with the help of (6.2.1). A simple example is now considered.

**Example 6.2.1** ***(A special pdf).*** Consider $f(x) = 4 \cdot (2+x)^{-2} - (1+x)^{-2}$ with cdf $F(x) = x^2/((1+x) \cdot (2+x))$. We shall try to find $X \sim \mathcal{T}$ such that, for some $n \geq 1$, f equals $f_n$ in (6.2.1). Trying $n = 1$ in (6.2.1) and substituting $s = x^{-1}$, we get

$$\phi_X'(s) = -s^{-2}f(1/s) = (1+s)^{-2} - 4\cdot(1+2s)^{-2}.$$

By an integration, $\phi_X(s) = \frac{1}{1+s}\cdot\frac{1}{1+2s} \in \check{\mathcal{T}}$. Hence $f \in \mathcal{T}$. □

We return to this example in Section 6.3. Theorem 6.2.1 can be generalized:

**Theorem 6.2.2.** *If $X$ and $Y$ are independent rv's and such that $X \sim \mathcal{T}$ and $Y - b \sim \mathcal{B}$ for some $b \geq 0$, then $X\cdot Y \sim \mathcal{T}$.*

**Proof.** By rescaling, we may assume that $b = 1$. Set $Z = Y-1$. Then

$$\phi_{XY}(s) = \phi_{X(Z+1)}(s) = \int \phi_X(s(y+1))f_Z(y)dy\,.$$

Hence and by the substitution $x = uv$, $y = u/v$,

$$\text{(6.2.2)}\qquad \phi_{XY}(st)\phi_{XY}(\tfrac{s}{t}) = \int 2u\left(\int \frac{1}{v}\,\phi_X(st(uv+1))\phi_X(\tfrac{s}{t}(\tfrac{u}{v}+1))\; f_Z(uv)f_Z(\tfrac{u}{v})\;dv\right)du$$

Now, $\phi_X(\cdot)\phi_X(\cdot)$ in the v–integral is a limit of products with factors of the form $\left((1 + cst(uv+1))\cdot(1 + c\frac{s}{t}(\frac{u}{v}+1))\right)^{-\kappa}$ $(c, \kappa > 0)$. By simple algebra we find that such a factor can be written as $\left(a_0 + a_1(t+t^{-1}) + a_2(v+v^{-1}) + a_3(tv+(tv)^{-1})\right)^{-\kappa}$, where $a_0, \ldots, a_3$ are positive constants depending on $s$ and $u$. This function is jointly CM in $t + t^{-1}$, $v + v^{-1}$, and $tv + (tv)^{-1}$ and therefore the product $\phi_X(\cdot)\phi_X(\cdot)$ can be represented as

$$\iiint \exp\{-\omega_1(t+\tfrac{1}{t}) - \omega_2(v+\tfrac{1}{v}) - \omega_3(tv+\tfrac{1}{tv})\}\;H(d\omega)\,,$$

where $H(d\omega) \geq 0$. In fact, $H(d\omega)$ is proportional to a trivariate GGC in the strong sense. Representing also $f_Z(uv)f_Z(u/v)$ as $\int \exp\{-\lambda u(v+v^{-1})\}K(d\lambda)$, we see, by a change of the order of integration between $v$ and $(\lambda,\omega)$ in (6.2.2) that $\phi_{XY}(st)\phi_{XY}(s/t)$ is CM in $t + t^{-1}$ for fixed $s > 0$ whenever

$$\int \frac{1}{v}\cdot\exp\{-\omega_1(t+\tfrac{1}{t}) - (\omega_2+\lambda u)(v+\tfrac{1}{v}) - \omega_3(tv+\tfrac{1}{tv})\}\;dv$$

is. However, the complete monotonicity in $t + t^{-1}$ of the latter integral was already verified in the proof of Theorem 5.1.1. Hence $\phi_{XY}$ is HCM, and thus, by Theorem 6.1.1, the proof is complete. □

A simple consequence of Theorem 6.2.2 appears.

**Theorem 6.2.3.** *If the distribution of* $X$ *is a denumerable convolution of Gamma distributions with shape parameters of size not less than 1, then* $e^X \sim \mathcal{T}$ *and, equivalently,* $e^X - 1 \sim \mathcal{T}$.

**Proof.** If $X_j \sim \text{Gamma}(\beta_j, \cdot)$, $\beta_j \geq 1$, then, by Corollary 2 in Section 5.7, $\exp\{X_j\} - 1 \sim \mathcal{B}$ and hence $\exp\{X_j\} \sim \mathcal{T}$. By Theorem 6.2.2 and induction, we get $\exp\{\Sigma_{j=1}^{n} X_j\} \sim \mathcal{T}$ when the rv's are independent. As $\mathcal{T}$ is closed with respect to weak limits, the first statement follows. The second one is then obvious since $e^X \geq 1$. □

In particular, if $X \sim PF_\infty$ (cf. Example 3.2.2), then $e^X \sim \mathcal{T}$. Thus the class of distributions corresponding to $e^X$, with $X \sim PF_\infty$, is a new simple subclass of $\mathcal{T}$ which is closed under multiplication of independent rv's.

It is not known whether the restriction on the size of the $\beta_j$'s can be removed. In a simple special case it can be removed; see Theorem 6.2.6. If $X \sim \mathcal{T} \Rightarrow X^q \sim \mathcal{T}$ (or, equivalently, $X \sim \mathcal{T} \Rightarrow (1+X)^q - 1 \sim \mathcal{T}$) for all $q \geq 1$ or for $q = 2$ at least, then Theorem 6.2.3 holds without any restriction. Theorem 6.2.3 will be applied in Chapter 8.

The next theorem is an improvement of Theorems 4.4.1 and 4.4.2. and illustrates the power of the real characterization in Theorem 6.1.1. As in Section 3.5, $\mathcal{T}_w$ is the class of widened GGC's and $\breve{\mathcal{T}}_w$ is the corresponding class of LT's (characterized by being decreasing and HCM, cf. Section 5.4.)

**Theorem 6.2.4.** *Let* $f \in \mathcal{T}$ *(or* $\mathcal{T}_w$*) and* $h \in \breve{\mathcal{T}}$ *(or* $\breve{\mathcal{T}}_w$*). Then* $h(x)f(x)$*, suitably normalized, belongs to* $\mathcal{T}$ *(or* $\mathcal{T}_w$*).*

**Remark.** Obviously $h(x)f(x)$ is the (defective) density for an rv stemming from $e^{-\theta x}f(x)$ with $\theta$ random with LT $h$; cf. Section 2.3. It is remarkable $C \cdot h \cdot f \in \mathcal{T}$ when $\theta \sim \mathcal{T}$ and $f \in \mathcal{T}$. □

**Proof.** It suffices to consider the case when $h(x) = (1+x)^{-\gamma}$, $\gamma > 0$. Now,

$$\psi(s) = \int e^{-sx}(1+x)^{-\gamma}f(x)dx = \frac{1}{\Gamma(\gamma)}\int y^{\gamma-1}e^{-y}\phi(s+y)dy ,$$

where $\phi$ is the LT of f. Hence

$$\psi(st)\psi(s/t) = \frac{1}{(\Gamma(\gamma))^2} \int\int x^{\gamma-1}y^{\gamma-1}e^{-x-y}\phi(st+x)\phi(\tfrac{s}{t}+y)dxdy .$$

Substituting for variation $x = u/v$, $y = uv$, we find by some algebra that, for fixed s and u, $\phi(st + u/v)\phi(s/t + uv)$ is a limit of products with factors of the form $\left(a_0 + a_1(t+t^{-1}) + a_2(v+v^{-1}) + a_3(tv+(tv)^{-1})\right)^{-\kappa}$, cf. the proof of Theorem 6.2.2. The rest of the proof proceeds as that proof. □

**Corollary.** *Let g be a nonnegative function on $(0,\infty)$ such that $\int e^{-xy}g(y)dy$ is HCM, (i.e., $g \in \mathcal{H}_W$) and let $\phi$ be decreasing and HCM. Then $\int g(y)\phi(s+y)dy$ ($= \int_s^\infty g(y-s)\phi(y)dy$) is HCM .*

**Proof.** When g itself is HCM, the statement follows by the proof of Theorem 6.2.4 with $y^{\gamma-1}e^{-y}$ replaced by $g(y)$. To prove the corollary in full, let $\phi(s) = \int e^{-sx}f(x)$, where $f \in \mathcal{H}_W$. Then

$$\int g(y)\phi(s+y)dy = \int g(y) \int e^{-(s+y)x}f(x)dx\, dy = \int e^{-sx}h(x)f(x)dx ,$$

where $h(x) = \int e^{-yx}g(y)dy$. Now the corollary follows from Theorem 6.2.4. □

A special case of the corollary was mentioned in Section 5.1; see (xi) there. Theorem 6.2.4 has many other consequences. Some (listed I – IV) are given below.

I) Letting f be a general Gamma density, we see that the class $\mathcal{B}$ again appears if weak limits are taken. Other subclasses of $\mathcal{T}$ with explicit densities appear if other densities f are chosen.

II) Choosing $h(x) = x^{\beta-1}$, $0 < \beta \le 1$ and f as in Theorem 4.2.1 we get (by simple algebraic manipulations) a generalization of that theorem:

**Theorem 6.2.5.** *A pdf f on $(0,\infty)$ of the form*

$$f(x) = \int \frac{1}{\Gamma(\beta)} x^{\beta-1} t^{\beta} e^{-xt} m(t)dt , \quad 0 < \beta \le 1,$$

*where $m(t)$ or (less restrictively) $g(t) = t^{\beta}m(t)$ is logconcave, is a GGC.*

III) We saw in Section 5.7 that if $X \sim \text{Gamma}(\beta, \lambda)$, $\beta \ge 1$, then $e^X - 1 \sim \mathcal{B}$. For $\beta < 1$ this result does not hold; cf. Section 5.8. However, still something can be said:

**Theorem 6.2.6.** *If $X$ is Gamma distributed, then $e^X - 1 \sim \mathcal{T}$. More generally, $(1+X)^q - 1 \sim \mathcal{T}$ for $q \geq 1$.*

**Proof.** Let $Y = e^X - 1$. Then $f_Y(x) = C\cdot(\log(1+x))^{\beta-1}(1+x)^{-1-\lambda}$. Consider the LT

$$\psi(s) = \int e^{-sx}(\log(1+x))^{\beta-1}\frac{1}{1+x}\,dx\,.$$

By an integration by parts,

$$\psi(s) = \beta^{-1}\cdot s \int e^{-sx}(\log(1+x))^{\beta}\,dx\,.$$

As $(\log(1+x))^{\beta}$ $(= x^{\beta}\cdot(\frac{\log(1+x)}{x})^{\beta})$ is HCM by Corollary 1 of Theorem 5.7.1, it follows that $\psi$ is HCM. Since moreover $\psi$ is decreasing, $(\log(1+x))^{\beta-1}\frac{1}{1+x} \in \mathcal{T}_W$. Hence, by Theorem 6.2.4 with $h(x) = (1+x)^{-\lambda}$, $f_Y \in \mathcal{T}$.

The proof of the second statement proceeds along the same lines. □

**IV)** Is it true that $X^q \sim \mathcal{T}$ for $q \geq 1$ when $X \sim \mathcal{T}$? The following result supports the suspicion that the answer could be positive.

**Theorem 6.2.7.** *If $X \sim PF_\infty$, then $X^q \sim \mathcal{T}$ for $q \geq 1$.*

**Proof.** It suffices to consider the case when the distribution of $X$ is a convolution of $n$ Exponential distributions with parameters $t_1 > t_2 > \ldots > t_n$. Let $f_n$ be the pdf of $X$ and set $\alpha = 1/q$. We use induction on $n$ to prove that the pdf $\alpha x^{\alpha-1}f_n(x^\alpha)$ of $X^q$ belongs to $\mathcal{T}$. We assume that $\alpha x^{\alpha-1}f_{n-1}(x^\alpha) \in \mathcal{T}$, whatever the values of $t_1, \ldots, t_{n-1}$ are. The assumption is true for $n = 2$. Now

$$f_n(x) = \int_0^x t_n\exp\{-t_n(x-y)\}\, f_{n-1}(y)dy = t_n e^{-t_n x}\int_0^x g_{n-1}(y)dy$$

where $g_{n-1}(y) = e^{t_n y}f_{n-1}(y)$. Here $g_{n-1}$ is proportional to the pdf of a convolution of $n-1$ Exponential densities with the parameters $t_1-t_n, t_2-t_n, \ldots, t_{n-1}-t_n$. By the induction hypothesis, $C\cdot x^{\alpha-1}g_{n-1}(x^\alpha) \in \mathcal{T}$. We get

$$\alpha x^{\alpha-1}f_n(x^\alpha) = t_n\alpha x^{\alpha-1}e^{-t_n x^\alpha}\int_0^{x^\alpha} g_{n-1}(y)dy = t_n\alpha x^{\alpha-1}e^{-t_n x^\alpha}\int_0^x \alpha z^{\alpha-1}g_{n-1}(z^\alpha)dz\,.$$

The integral is essentially the cdf of a GGC and therefore a function in $\mathcal{T}_W$. The factor in front of the integral is HCM and decreasing. Thus, by Theorem 6.2.4,

$\alpha x^{\alpha-1} f_n(x^\alpha) \in \mathcal{T}$. □

Some generalizations of Theorems 6.2.3, 6.2.6 and 6.2.7 are possible but the conjecture that $X \sim \mathcal{T} \Rightarrow X^q \sim \mathcal{T}$ for $q \geq 1$ remains unproved.

## 6.3 Some Nested Subclasses of $\mathcal{T}$ and Their Properties

In this section we shall look at two kinds of nested subclasses of $\mathcal{T}$, the $\mathcal{R}_k$–classes and the $\mathcal{T}^k$–classes. These classes can be described by HCM–conditions. Some classes which are not included in $\mathcal{T}$, but are defined by HCM–conditions, are also considered.

***The ratio classes $\mathcal{R}_k$.*** Let $X \sim \mathcal{T}$ $(X > 0)$ and $Y_k \sim \text{Gamma}(k,1)$, $0 < k \in \mathbb{R}$, be independent rv's. Then, as was found in Section 6.2, $X/Y_k \sim \mathcal{T}$. We have

$$f_{X/Y_k}(x) = \frac{1}{\Gamma(k)} x^{-(k+1)} \int \exp\{-\tfrac{y}{x}\} y^k F_X(dy) . \tag{6.3.1}$$

Let $\mathcal{R}_k$ denote the class of such pdf's; $\mathcal{R}_k \subset \mathcal{T}$. These classes $\mathcal{R}_k$, $k > 0$, have some interesting properties. Of course, the weak closure of $\lim_{k\to\infty} \mathcal{R}_k$ equals $\mathcal{T}$.

**Theorem 6.3.1.** *For $0 < j < k$ (j and k real), $\mathcal{R}_j \subset \mathcal{R}_k$ .*

**Proof.** Consider $X/Y_j$ with $X \sim \mathcal{T}$. Now $Y_j = Z \cdot Y_k$, where $Z$ and $Y_k$ are independent and $Z \sim \text{Beta}(j, k-j)$. We have $X/Y_j = (X \cdot Z^{-1})/Y_k$. Thus it suffices to verify that $X \cdot Z^{-1} \sim \mathcal{T}$. The pdf of $Z^{-1} - 1$ equals $C \cdot x^{k-j-1}/(1+x)^k$ and belongs to $\mathcal{B}$. Theorem 6.2.2 then gives the desired result. □

Let $f$ be any pdf and set

$$F^{[k]}(x) = \frac{1}{\Gamma(k)} \int_0^x (x-y)^{k-1} f(y)dy \qquad (k > 0).$$

As is well known and easily verified,

$$F^{[k]}(x) = \frac{1}{\Gamma(k-j)} \int_0^x (x-y)^{k-j-1} F^{[j]}(y)dy .$$

The transform $F^{[k]}$ is an Euler tranform of the first kind and determines $f$ uniquely. In fact, the LT of $F^{[k]}(x)$ equals $s^{-k}\phi(s)$, where $\phi$ is the LT of f. Obviously $F^{[1]}$ is the cdf $F$ of $f$ and $F^{[2]}(x) = \int_0^x F(y)dy$. Moreover, $F^{[k]}(x) \to f(x)$ as $k \to 0$. There is a simple characterization of the class $\mathcal{R}_k$ in terms of $F^{[k]}$.

**Theorem 6.3.2.** *A pdf $f$ belongs to $\mathcal{R}_k$ (and hence to $\mathcal{T}$) iff $F^{[k]}$ is HCM.*

To prove the theorem a lemma is needed.

**Lemma.** *For independent variables $X > 0$ and $Y_k \sim Gamma(k,1)$,*

$$F^{[k]}_{X/Y_k}(x) = \frac{1}{\Gamma(k)} x^{k-1} \int e^{-y/x} F_X(dy) = \frac{1}{\Gamma(k)} x^{k-1} \phi_X(\tfrac{1}{x}).$$

**Proof.** The pdf of $X/Y_k$ is given by (6.3.1). Hence

$$F^{[k]}_{X/Y_k}(x) = \frac{1}{\Gamma(k)} \int_0^x (x-z)^{k-1} f_{X/Y_k}(z)dz = \frac{1}{(\Gamma(k))^2} \int y^k \Big( \int_0^x (x-z)^{k-1} z^{-k-1} e^{-y/z} dz \Big) F_X(dy)$$

Substituting $z = x/t$ in the z–integral, we find that it equals

$$\frac{1}{x} \int_1^\infty (t-1)^{k-1} \exp\{-\tfrac{y}{x} t\} dt = \Gamma(k)\, x^{k-1} y^{-k} e^{-y/x};$$

this gives the desired result. □

**Proof of Theorem 6.3.2.** The "only if" part is obvious from the Lemma; $x^{k-1}\phi_X(\frac{1}{x})$ is HCM if $X \sim \mathcal{T}$. Turning to the "if" part, we first note that if $F^{[k]}$ is HCM, then so is its LT $s^{-k}\phi(s)$, where $\phi(s) = \int e^{-sy} f(y)dy$. Hence $\phi$ is HCM and thus $f \in \mathcal{T}$ by Theorem 6.1.1. By the unimodality of $f$, $xf(x) \to 0$ as $x \to \infty$. Now, since $\int f(y)dy = 1$,

$$g(x) = \Gamma(k) x^{-k+1} F^{[k]}(x) = \int_0^x (1 - \tfrac{y}{x})^{k-1} f(y)dy \to 1$$

as $x \to \infty$. This result is obvious for $k \geq 1$ by dominated convergence. Splitting the integral into two parts and using that $yf(y) \to 0$ as $y \to \infty$, we also easily find that it holds for $k < 1$. Since now also $g(x^{-1})$ is HCM and has value $1$ at $x = 0$, it is the LT of a GGC by Theorem 6.1.1. Hence, with $X \sim \mathcal{T}$,

$$F^{[k]}(x) = \frac{x^{k-1}}{\Gamma(k)} \int e^{-y/x} F_X(dy) = F^{[k]}_{X/Y_k}(x),$$

where the last equality is a consequence of the lemma. By the uniqueness of the transform $F^{[k]}$, it follows that $f(x) = f_{X/Y_k}(x)$, i.e., $f \in \mathcal{R}_k$. □

The most interesting particular case of Theorem 6.3.2 is of course $k = 1$:

*If the cdf* $F$ *is HCM, or, equivalently, if* $F(x^{-1})$ *is the LT of a GGC, then* $f \in \mathcal{R}_1$ *and hence* $f \in \mathcal{G}$. Recalling Example 6.2.1, we see that there $F(x) = x^2/((1+x)\cdot(2+x))$ and hence $F(x^{-1}) = (1+x)^{-1}\cdot(1+2x)^{-1} \in \check{\mathcal{G}}$. It is easy to show that $X \sim \mathcal{R}_1 \Rightarrow X^q \sim \mathcal{R}_1$ (and hence $X^q \sim \mathcal{G}$) for $q \geq 1$. Since $F^p$ is HCM for $p > 0$ whenever $F$ is, the class $\mathcal{R}_1$ is closed with respect to formation of maxima of independent rv's and all distributions in $\mathcal{R}_1$ are ID also with respect to the max–operation. Moreover, with respect to that operation, $\mathcal{R}_1$ is generated by the cdf's $F(x) = C\cdot x^\gamma/(1+cx)^\gamma$, $\gamma > 0$, $c > 0$, corresponding to ratios $Z/Y$, where $Z \sim \text{Gamma}(\gamma,c)$ and $Y \sim \text{Exp}(1)$ are independent.

**Remark 1.** Let $\mathcal{R}_0 = \lim_{k\to 0} \mathcal{R}_k$. As $\lim_{k\to 0} F^{[k]}(x) = f(x)$, it follows that $\mathcal{R}_0 \subset \mathcal{B}$. However, it is not true that $\mathcal{R}_0 = \mathcal{B}$. In fact, $\mathcal{R}_0$ can be verified to be empty! □

**Remark 2.** The classes $\mathcal{R}_k$ can be enlarged. By Theorem 6.2.4, every pdf of the form $C\cdot h(x)f(x)$, where $f \in \mathcal{R}_k$ and $h \in \check{\mathcal{S}}_w$, belongs to $\mathcal{G}$. In particular, we may consider the class $\tilde{\mathcal{R}}_k$ of pdf's of the form $C\cdot x^{-\gamma}f(x)$, $f \in \mathcal{R}_k$, $\gamma \geq 0$. Obviously $\tilde{\mathcal{R}}_j \subset \tilde{\mathcal{R}}_k \subset \mathcal{G}$ for $j < k$. It is not hard to see from (6.3.1) that the weak closure of $\tilde{\mathcal{R}}_0 = \lim_{k\to 0} \tilde{\mathcal{R}}_k$ essentially equals $\mathcal{B}$. Obviously, $X \sim \tilde{\mathcal{R}}_1 \Rightarrow X^q \sim \tilde{\mathcal{R}}_1$ for $q \geq 1$. □

***The*** $\mathcal{G}^k$***–classes.*** Let us now consider the classes $\mathcal{G}^k$, $k \in \mathbb{R}_+$, of pdf's $f$ such that $C\cdot x^k e^{-\epsilon x} f(x) \in \mathcal{G}$ for all $\epsilon > 0$. They have relations to the classes $\mathcal{R}_k$ when multiplication of rv's is considered. Theorem 6.2.4 shows that $\mathcal{G}^j \supset \mathcal{G}^k$ for $j < k$. Further, $\mathcal{G}^0 = \mathcal{G}$ (essentially) and $\mathcal{G}^\infty = \lim_{k\to\infty} \mathcal{G}^k = \mathcal{B}$. The following theorem is just a variant of Theorem 6.1.2.

**Theorem 6.3.3.** *A pdf* $f$ *belongs to* $\mathcal{G}^k$ *iff* $\int e^{-xy} y^k f(y)dy$ *is HCM.*

An simple consequence of this theorem and (6.3.1) is the following characterization.

**Theorem 6.3.4.** *Let* $X$ *be a non–degenerate rv. Then* $X \sim \mathcal{G}^k$, $k > 0$, *iff* $X/Y_k \sim \mathcal{B}$. *(As before* $Y_k$ *is independent of* $X$ *and* $Y_k \sim \text{Gamma}(k,1)$.*)*

In view of Theorems 5.1.1 and 6.2.1, we obtain the following corollaries which can be considered to be generalizations of these theorems.

**Corollary 1.** *If* $X \sim \mathcal{G}^k$ *and* $Y \sim \mathcal{B}$ *(*$X$ *and* $Y$ *independent), then* $X\cdot Y \sim \mathcal{G}^k$.

**Corollary 2.** *If* $X_1 \sim \mathcal{G}^k$ *and* $X_2 \sim \mathcal{R}_k$ *(*$X_1$ *and* $X_2$ *independent), then* $X_1\cdot X_2 \sim \mathcal{G}$.

*The $\mathscr{A}_k$–classes.* For completeness we end this section by studying also the pdf's f for which the Euler transform of the second kind

$$F^{[k]}(x) = \frac{1}{\Gamma(k)} \int_x^\infty (y-x)^{k-1} f(y)dy\,, \quad k > 0,\ k \in \mathbb{R},$$

is HCM. The class of such pdf's is denoted $\mathscr{A}_k$.

**Theorem 6.3.5.** *If $f \in \mathscr{A}_k$, then $f$ is CM (and hence ID). More precisely, $f \in \mathscr{A}_k$ iff $f(x) = \int t \cdot e^{-xt} M(dt)$, where $t^{-k+1} M(dt) \in \mathscr{T}_W$. Moreover, $\mathscr{A}_j \subset \mathscr{A}_k$ for $j < k$.*

**Proof.** Let $f \in \mathscr{A}_k$. Since $F^{[k]}(x)$ is HCM and has the limit $\int y^{k-1} f(y)dy > 0$ as $x \to 0$, it follows that $F^{[k]}$ is *decreasing* and CM; cf. Section 5.4. Differentiating $F^{[k]}$ m times, where the integer m is chosen so that $0 \leq k-m < 1$, we find that $F^{[k-m]}$ is also CM. Then replacing k by 1 and j by k–m in the well–known formula

$$F^{[k]}(x) = \frac{1}{\Gamma(k-j)} \int_x^\infty (y-x)^{k-j-1} F^{[j]}(y)dy \qquad (j < k), \tag{6.3.2}$$

we easily find that $\overline{F}(x)$ is CM. Hence f is also CM. Representing f as $f(x) = \int te^{-xt} M(dt)$, we find, by some calculation, that $F^{[k]}(x) = \int e^{-xt} t^{-k+1} M(dt)$. Hence $f \in \mathscr{A}_k$ iff $t^{-k+1} M(dt) \in \mathscr{T}_W$.

Theorem 6.2.4 shows that $\mathscr{A}_j \subset \mathscr{A}_k$ for $j < k$. The latter result follows also from the corollary of that theorem in conjunction with (6.3.2). □

The class $\mathscr{A}_1$ equals the class of pdf's for which the survival function $\overline{F}(x) = 1 - F(x)$ is the LT of a GGC and is particularly interesting. It is the smallest class that is closed with respect to weak limits and formation of minima of independent rv's and that contains the Pareto distributions: $\overline{F}(x) = (1+cx)^{-\gamma}$, $\gamma > 0$. Obviously $X \sim \mathscr{A}_1$ iff $X \overset{d}{=} Y/\theta$, where $Y \sim \text{Exp}(1)$ and $\theta \sim \mathscr{T}$ are independent rv's. Hence $X \sim \mathscr{A}_1$ iff $1/X \sim \mathscr{R}_1$.

It is apparent that $\mathscr{A}_0 = \lim_{k \to 0} \mathscr{A}_k$ equals the class of decreasing pdf's in $\mathscr{B}$ and is closed with respect to multiplication of densities. The class $\mathscr{A}_0$ rather trivially shares some other properties with the class $\mathscr{B}$:

*i)* *If $X \sim \mathscr{A}_0$, then $X^q \sim \mathscr{A}_0$ for $q \geq 1$.*

*ii)* *If $X \sim \mathscr{A}_0$ and $Y \sim \mathscr{A}_0$ are independent, then $X \cdot Y \sim \mathscr{A}_0$ and $X/Y \sim \mathscr{A}_0$.*

The larger classes of CM and logconvex densities also have these properties. If f is CM (or logconvex), then so is the survival function $\overline{F}$. Logconvexity of $\overline{F}$ means that the *failure rate* $f(x)/\overline{F}(x)$ is decreasing; cf., e.g., Barlow & Proschan (1975, p. 76). For

$f \in \mathscr{S}_1$, and thus also for $f \in \mathscr{S}_0$, the failure rate is the Stieltjes transform of a nonnegative measure and thus CM, in particular. □

## 6.4 Summary and a New Perspective

In Chapters 5 and 6 the HCM–condition has been applied and studied intensively. Classes of distributions have been identified for which pdf's (f), Laplace transforms ($\phi$), and Euler transforms are HCM. The following table shows (except for some gaps) whether or not a class is closed with respect to some operations on independent rv's. Most of the results have been discussed above. A few further comments are given.

**Table 6.4.1.** *Preservation of various classes under some operations (given in terms of operations on rv's). Small numbers indicate that comments follow.*

| Class | Definition | Operations on rv's: $X_1+X_2$ | $X_1 \cdot X_2$ | $X^q$ | $e^X-1$ |
|---|---|---|---|---|---|
| $\mathscr{B}$ | f HCM | No | Yes | Y if $\lvert q\rvert \geq 1$ | N |
| $\mathscr{S}_0$ | f HCM, f↓ | N | Y | Y if $q \geq 1$ | N |
| $\mathscr{R}_1$ | F HCM | N | [1] ? | Y " " | [2] N |
| $\mathscr{S}_1$ | 1−F HCM | N | [1] ? | Y " " | [2] N |
| $\mathscr{T}$ | $\phi$ HCM | Y | [1] ? | ? " " | [1] ? |
| $PF_\infty$ | | Y | N | N; $X^q \sim \mathscr{T}$, $q \geq 1$ | N; $e^X-1 \sim \mathscr{T}$ |

Comments:

1) There are indications (cf. Table 6.4.2 below) that not only $\mathscr{B} \cdot \mathscr{R}_1 \subset \mathscr{R}_1$ but also $\mathscr{R}_1 \cdot \mathscr{R}_1 \subset \mathscr{R}_1$ (and hence $\mathscr{S}_1 \cdot \mathscr{S}_1 \subset \mathscr{S}_1$). It can be shown that then $\mathscr{R}_1 \cdot \mathscr{T} \subset \mathscr{T}$. Moreover, some numerical experiments indicate that $PF_\infty \cdot \mathscr{T} \subsetneq \mathscr{T}$. If, at best, $\mathscr{T} \cdot \mathscr{T} \subsetneq \mathscr{T}$, then it would trivially follow that $X \sim \mathscr{T} \Rightarrow e^X-1 \sim \mathscr{T}$ (since $e^Y \sim \mathscr{T}$ if $Y \sim$ Gamma).

2) By tedious complex analysis or by numerical methods it can be shown that $F(x) = x/(1+x)$ and $\overline{F}(x) = (1+x)^{-1}$ are not HCM when x is replaced by $\log(1+x)$.

The HCM–condition may look exotic. It is therefore informative to relate it to something better known. Obviously it has connections to complete monotonicity and logconvexity. However, it is most strongly related to logconcavity. Consider the functions $f\ (\geq 0)$ for which it is *only* assumed that $f(uv)f(u/v)$ is decreasing in $w = v + v^{-1}$. Here such a function is called *hyperbolically monotone* (HM). Most of the properties listed in Section 5.1 for HCM–functions also hold for HM–functions; a few do not. As essentially already found in Section 5.1, we have:

*A measurable function $f$ ($\geq 0$) is HM iff $f(e^y)$ (or $e^y f(e^y)$) is logconcave.*

(The "if" part is rather obvious from the geometric meaning of logconcavity.)

Thus the pdf of an rv $X > 0$ is HM iff $\log X$ has a logconcave (i.e. strongly unimodal or $PF_2$–) density. The survival function of $X$ is HM iff $\log X$ has a logconcave survival function, i.e. a distribution on $(-\infty, \infty)$ with *increasing failure rate* (IFR). The cdf of $X$ is HM iff $-\log X \sim$ IFR. Using inter alia the known properties of the classes of life distributions (see, e.g., Barlow & Proschan (1975, Chapter 4)), we get the following table for the HM–classes corresponding to the HCM–classes above.

**Table 6.4.2.** *Preservation of various HM–classes under some operations. As to the inclusions, SD means the class of self–decomposable distributions.*

| Class | Definition | Inclusions | Operations on rv's | | | |
|---|---|---|---|---|---|---|
| | | | $X_1+X_2$ | $X_1 \cdot X_2$ | $X^q$ | $e^X$ |
| $\mathcal{B}$ | f HM | $\mathcal{B} \subset \mathcal{T}$ | No | Yes | Yes | N |
| $\mathcal{S}_0$ | f HM, f↓ | $\mathcal{S}_0 \subset \mathcal{B}$ | N | Y | Y if $q \geq 0$ | N |
| $\mathcal{R}_1$ | F HM | $\mathcal{R}_1 \subset \mathcal{T}$ | N | Y | Y " " | Y |
| $\mathcal{S}_1$ | 1–F HM | $\mathcal{S}_0 \subset \mathcal{S}_1$ | N | Y | Y " " | N |
| $\mathcal{T}$ | $\phi$ HM | $\mathcal{T} \supset$ SD | Y | ? | ? " " | ? |

Only a few comments are given concerning the new classes, whose symbols are chosen to resemble those of the HCM–classes, and the table above.

1) It is well known that the $PF_2$– and IFR–classes are closed with respect to convolution. These results give that $\mathcal{B}$, $\mathcal{S}_0$, $\mathcal{R}_1$, and $\mathcal{S}_1$ are closed with respect to multiplication of independent rv's. (The proof of Theorem 5.1.1 can also be easily modified to show that $\mathcal{B}$ has this property. The function $g(uv)g(u/v)$ there should be approximated by a decreasing piecewise constant function of $w$.)

2) The new class $\mathcal{T}$ is large. It can be shown to contain all SD–distributions. However, as $\mathcal{T} \supset \mathcal{B}$, it also contains distributions which are not ID, e.g. a Beta($\alpha,\beta$)–distribution with $\beta \geq 1$. It can be shown that if $C \cdot x^n e^{-\epsilon x} f(x) \in \mathcal{T}$ for all $\epsilon > 0$ and $n \in \mathbb{N}_0$, then $f \in \mathcal{B}$. However, few of the many other specific results for the HCM–class $\mathcal{G}$ admit generalizations to the HM–class $\mathcal{T}$.

We end this chapter by giving the pdf of $\log X$ for $X \sim \mathcal{B}$. In the simplest case the pdf has the form

$$f_{\log X}(x) = C \cdot e^{\beta x} \cdot \prod_{i=1}^{n} \left(\cosh(x-\theta_i)\right)^{-\gamma_i}, \quad x \in \mathbb{R},$$

with $\theta_i \in \mathbb{R}$, $\gamma_i > 0$, and $|\beta| < \Sigma\, \gamma_i$. It is not easy to see that the class of pdf's of this form and limits thereof is closed with respect to convolution. These distributions may be considered as broad generalizations of the Logistic distribution; cf. Example 7.2.4.

### 6.5 Complementary Results and Notes

I (***On LT's which are HCM***). An LT which is HCM does not always correspond to a (widened) GGC. Counterexample: $\int_0^\infty e^{-sx}\sin(\sqrt{x})dx = (\sqrt{\pi}/2)\, s^{-3/2}\exp\{-1/(4s)\}$; cf. (3.1.9).

II (***A reformulation***). Theorem 6.1.1 has the following variant: A probability distribution on $\mathbb{R}_+$ is a GGC iff its LT $\phi$ has the property that

$$-\frac{s\phi'(s)\phi(t) - t\phi'(t)\phi(s)}{s - t}$$

is CM in $t+s$ when $t\cdot s$ is held fixed. In this formulation, Theorem 6.1.1 resembles Theorem 3.4.2 (expressed for mgf's). However, it seems very difficult to prove that these two theorems are equivalent.

III (***Two counterexamples***). Let $f$ be a pdf on $(0,\infty)$. If $-f'(x)/f(x)$ is the Stieltjes transform of a nonnegative measure, then so is $f(x)/\overline{F}(x)$. This is a reformulation of the fact that $\mathscr{A}_0 \subset \mathscr{A}_1$; cf. Section 6.3. One may ask whether complete monotonicity of $-f'(x)/f(x)$ implies that $f(x)/\overline{F}(x)$ is CM. However, by tedious calculations it can be verified that $f(x) = \exp\{-2x+e^{-x}-1\}$ (with $\overline{F}(x) = 1-(1-e^{-x})\cdot\exp\{e^{-x}\}$ ) works as a counterexample.

We have seen that $f \in \mathscr{T} \Rightarrow C\cdot x^{-1}f(x) \in \mathscr{T}$. This implication holds also if $\mathscr{T}$ is replaced by the class of CM pdf's. Can $\mathscr{T}$ be replaced by the class of all ID distributions? It cannot! In fact, the LT $\phi(s) = \exp\{-2s+e^{-s}-1\}$, corresponding to a translated Poisson distribution, is ID, i.e., $-\phi'/\phi$ is CM, but the LT $\psi(s) = \int_s^\infty \phi(t)dt$, induced by the multiplication by $x^{-1}$, is not ID since $-\psi'/\psi$ is not CM, as mentioned above.

IV (***An open problem***). As mentioned in Section 2.1, $f\star g$ is unimodal whenever $g$ is iff $f \in PF_2$. It is not known whether there is a similar characterization of the class $\mathscr{B}$ with respect to multiplicative convolution. In view of Theorem 6.2.2, it is not true that $X \sim \mathscr{T} \Rightarrow Y\cdot X \sim \mathscr{T}$ iff $Y \sim \mathscr{B}$.

V (***The $HM_k$-classes***). A nonnegative function $f$ is said to be *hyperbolically monotone of order $k$* ($HM_k$) if, for every $u > 0$, the function $g(w) = f(uv)f(u/v)$, $w = v + v^{-1}$, is such that $(-1)^j g^{(j)}(w) \geq 0$, $j = 1, 2, \ldots, k$. We know that for $k = 1$ and $k = \infty$ the $HM_k$–property is preserved under multiplicative convolution. It can be shown to be preserved also for $k = 2$. Numerical computations indicate that it is preserved for all $k \in \mathbb{N}$, but no proof exists.

**Notes.** Many of the results in this chapter are new. Some of the results appeared in B1990.

# 7. EXTENDED GENERALIZED GAMMA CONVOLUTIONS

Many distributions are limit distributions for sums of independent positive and *negative* Gamma variables. According to the central limit theorem, the N(0,1)–distribution with mgf $\exp\{s^2/2\}$ is a simple example of such a distribution. An extension of $\mathscr{T}$ to a class of distributions on the whole real line $\mathbb{R}$ is thus desirable. This chapter is devoted to the class $\mathscr{T}_e$ of extended generalized Gamma convolutions (EGGC) introduced by Thorin (1978a).

In Section 7.1 the EGGC's are defined and studied via the mgf. Several examples of EGGC's of the parade type are given in Section 7.2. In Section 7.3 variance mixtures of the Normal distribution are considered, and the real line counterpart of the $\mathscr{B}$–class is introduced. This new class, $\mathscr{B}_e$, also has some multiplicative properties. Section 7.4 treats composition of mgf's. In Section 7.5 self–reciprocal distributions are studied.

## 7.1 Definition and Basic Theory

It is most convenient to start by the formal definition.

**Definition.** *An extended generalized Gamma convolution (EGGC) (distribution in $\mathscr{T}_e$) is a probability distribution on $\mathbb{R}$ such that the mgf $\varphi(s) = \int e^{sx}F(dx)$, defined at least for $\Re e\ s = 0$, has the form*

$$(7.1.1) \qquad \varphi(s) = \exp\{bs + \frac{cs^2}{2} + \int \left(\log(\frac{t}{t-s}) - \frac{st}{1+t^2}\right) U(dt)\}\ ,$$

*where $b \in \mathbb{R}$, $c \geq 0$, and $U(dt)$ is a nonnegative measure on $\mathbb{R}\backslash\{0\}$ such that*

$$(7.1.2) \qquad \int \frac{1}{1+t^2}\ U(dt) < \infty \quad \text{and} \quad \int_{|t|\leq 1} |\log t^2|\ U(dt) < \infty\ .$$

Integrals in this chapter are calculated over $\mathbb{R}\backslash\{0\}$ or $\mathbb{R}$ unless otherwise indicated. The term $st/(1+t^2)$ is necessary to ensure the convergence of the integral. It can be omitted if $\int_{|t|>1}|t|^{-1}U(dt) < \infty$. Hence $\mathscr{T} \subset \mathscr{T}_e$. Obviously a finite convolution of Gamma distributions on the positive or negative real axis is an EGGC. Note that the mgf of a Gamma distribution on $(-\infty,0)$ has the form $(\frac{t}{t-s})^u$ with $t < 0$. Moreover, $\mathscr{T}_e$ equals the class of all possible limit distributions for such finite convolutions and is closed with respect to weak limits. This is a consequence of the closure theorem which is formulated and proved below. The class is also closed with respect to scale and location changes of the distributions.

Obviously the mgf of an EGGC can be extended to a function which is analytic in $\mathbb{C}\setminus\mathbb{R}$. The whole real line is a cut. We have $\varphi(\bar{s}) = \overline{\varphi(s)}$ in accordance with the relation $\varphi(-is) = \overline{\varphi(is)}$, $s \in \mathbb{R}$, valid for characteristic functions (chf). By differentiation,

$$\frac{\varphi'(s)}{\varphi(s)} = b + cs + \int \left(\frac{1}{t-s} - \frac{t}{1+t^2}\right) U(dt) . \tag{7.1.3}$$

This is the general form of a Pick function; cf. Section 2.4, and it follows that b, c, and U(dt) are uniquely determined by $\varphi$. If $\varphi$ is a function that satisfies (7.1.3) and $\varphi(0) = 1$, then an integration of (7.1.3) leads back to (7.1.1) with the conditions (7.1.2) and $U(\{0\}) = 0$.

We now formulate and sketch a proof of the closure theorem for $\mathcal{T}_e$. It is appropriate to use a modified representation of the mgf $\varphi$.

Let $\bar{\mathbb{R}}$ be the compactified real line with a single infinity point $\infty$. Let further

$$g(s,t) = \frac{\log\left(\frac{t}{t-s}\right) - \frac{st}{1+t^2}}{\log(1+t^{-2})}, \quad s \in \mathbb{C}\setminus\mathbb{R} ; \quad g(0,t) \equiv 0.$$

The function $t \to g(s,t)$ is continuous on $\bar{\mathbb{R}}$ if we define, for $s \neq 0$, $g(s,0) = -1/2$ and $g(s,\infty) = s^2/2$. The mgf of an EGGC can now in a unique way be represented as

$$\varphi(s) = \exp\{bs + \int_{\bar{\mathbb{R}}} g(s,t)\nu(dt)\},$$

where $\nu$ is a nonnegative measure on $\bar{\mathbb{R}}$ such that $\nu(\{0\}) = 0$. In fact, $\nu(dt) = \log(1+t^{-2})U(dt)$, $t \in \mathbb{R}\setminus\{0\}$; $\nu(\{\infty\}) = c$.

**Theorem 7.1.1** ***(Closure theorem).*** *Let $\{F_n\}_1^\infty$ be a sequence of EGGC's given by $\{(b_n,\nu_n)\}_1^\infty$. If $F_n \to F$ weakly, then $F \in \mathcal{T}_e$. Moreover, if $F$ is given by $(b,\nu)$, then*

$$b_n \to b \quad \text{and} \quad \nu_n \to \nu \quad \text{vaguely on } \bar{\mathbb{R}} . \tag{7.1.4}$$

*Conversely, if (7.1.4) holds for some $b \in \mathbb{R}$ and $\nu$ such that $\nu(\{0\}) = 0$, then $F_n \to F$ weakly with $F \in \mathcal{T}_e$ defined by $(b,\nu)$.*

**Sketch of Proof.** We must use the continuity theorem for chf's stating that $F_n \to F$ iff $\varphi_n(is) \to \varphi(is)$, $s \in \mathbb{R}$. Let $F_n \to F$ weakly. Then, in particular, $|\varphi_n(is)|^2 \to |\varphi(is)|^2$. Now

$$|\varphi_n(is)|^2 = \exp\left\{ \int_{\bar{\mathbb{R}}} \frac{\log\left(\frac{t^2}{t^2+s^2}\right)}{\log(1+t^{-2})} \nu_n(dt) \right\} .$$

The integrand is negative with maximum value $-\min(1, s^2)$. Since $|\varphi(is)| > 0$ at least for some $s \neq 0$ close to zero, it follows that the sequence $\{\int_{\mathbb{R}} \nu_n(dt)\}_1^\infty$ is bounded. Thus, any subsequence of $\{\nu_n\}_1^\infty$ can be thinned to a sequence $\{\nu_{n_j}\}_{j=1}^\infty$ that tends vaguely on $\overline{\mathbb{R}}$ to some $\nu$ as $j \to \infty$. It follows that, for all $s \in \mathbb{R}$,

$$\exp\{\int_{\mathbb{R}} g(is,t)\nu_{n_j}(dt)\} \to \exp\{\int_{\mathbb{R}} g(is,t)\nu(dt)\}, \quad j \to \infty.$$

Hence $\exp\{ib_{n_j}s\}$ also converges for all $s \in \mathbb{R}$. It is known that then $b_{n_j}$ has a limit $b \in \mathbb{R}$. Hence

$$\varphi(is) = \exp\{ibs + \int_{\mathbb{R}} g(is,t)\nu(dt)\}.$$

Since $\varphi(0) = 1$ we have $\nu(\{0\}) = 0$. Thus $F \in \mathcal{T}_e$ and $b$ and $\nu$ are uniquely determined. This uniqueness then implies that (7.1.4) holds.

The converse part is obvious. □

Consider now an EGGC given by $(b,\nu)$ with $\nu(\{0\}) = 0$. We can choose measures $\nu_n$, $n = 1, 2, \ldots$, each with mass only at a finite number of points in $\mathbb{R}\backslash\{0\}$ and such that $\nu_n \to \nu$ vaguely on $\overline{\mathbb{R}}$. Setting $b_n \equiv b$, we see that the EGGC is a weak limit of a sequence of translated finite convolutions of Gamma distributions on $(-\infty,0)$ or $(0,\infty)$. The convolution corresponding to $n$ is translated

$$b - \int \frac{t}{1+t^2} U_n(dt), \quad \text{where} \quad U_n(dt) = \frac{1}{\log(1+t^{-2})} \nu_n(dt),$$

units to the right. Then, modifying $\nu_n$ slightly by giving $U_n(dt)$ an additional mass at a point sufficiently far from the origin, we can arrange so that the translation is zero for each $n$ without affecting the limit measure $\nu$. Thus any EGGC is a limit of finite convolutions of Gamma distributions.

An EGGC is self–decomposable and it is easily verified that the Lévy measure has a density $\ell$ such that

$$(7.1.5) \qquad |y|\ell(y) = \begin{cases} \int_{(0,\infty)} e^{-yt}U(dt), & y > 0 \\ \int_{(-\infty,0)} e^{yt}U(dt), & y < 0 \end{cases}$$

Because of the self–decomposability, every non–degenerate EGGC is absolutely continuous and unimodal; cf. Section 2.3.

The EGGC's can be characterized in the same way as the GGC's. A distribution on $\mathbb{R}$ belongs to $\mathcal{T}_e$ iff its mgf $\varphi(s)$ (defined at least for $\Re e\, s = 0$) can be analytically continued to $\mathbb{C}\backslash\mathbb{R}$ and is zero–free there and $\Im m[\varphi'(s)/\varphi(s)] \geq 0$ for $\Im m\, s > 0$. This was essentially noticed already in the beginning of this section. The inversion formula for Pick functions gives

$$\text{(7.1.6)} \qquad c = \lim_{\eta\uparrow\infty} \frac{1}{i\eta}\frac{\varphi'(i\eta)}{\varphi(i\eta)} \quad \text{and} \quad \int_{(d,e)} U(dt) = \lim_{\eta\downarrow 0} \frac{1}{\pi} \int_{(d,e)} \Im m\left[\frac{\varphi'(x+i\eta)}{\varphi(x+i\eta)}\right] dx$$

if $U(\{d\}) = U(\{e\}) = 0$. The characterization in Theorem 3.1.3 of GGC's can be immediately extended to a powerful characterization of EGGC's.

**Theorem 7.1.2.** *A distribution on* $\mathbb{R}$ *is an EGGC iff its mgf* $\varphi$ *can be analytically continued to the open upper half–plane and* $\Im m[\varphi'(s)\overline{\varphi(s)}] \geq 0$ *there.*

**Remark.** Even a function $\varphi$, satisfying $\varphi(0) = 1$ and the conditions above, is the mgf of an EGGC. □

The composition theorem for GGC's in Section 3.3 also has a generalization. Let $\hat{\mathscr{T}_e}$ denote the class of mgf's of EGGC's.

**Theorem 7.1.3** ***(Composition theorem).*** *Let* $e^{\zeta(s)} \in \hat{\mathscr{T}_e}$ *(with* $\zeta(0) = 0$ *and* $\zeta$ *continuous). Then* $\varphi(\zeta(s)) \in \hat{\mathscr{T}_e}$ *whenever* $\varphi \in \hat{\mathscr{T}}$ *iff* $\Im m[\zeta'(s)\overline{\zeta(s)}] \leq 0$ *for* $\Im m\, s > 0$.

It should be recalled that if $X(\tau)$, $\tau > 0$, is a Lévy process corresponding to $e^{\tau\zeta(s)}$ and $T \geq 0$ is an independent rv with mgf $\varphi$, then $\varphi(\zeta(s))$ is the mgf of $X(T)$. The proof of Theorem 7.1.3 is not given. It follows completely the lines of the proof of Theorem 3.3.1 if the following lemma, which is of independent interest, is used. A proof which does not rely on the lemma can also be given; cf. the remark following the proof of the lemma.

**Lemma.** *Let* $e^{\zeta(s)} \in \hat{\mathscr{T}_e}$, *where* $\zeta(s) \not\equiv 0$. *Then* $\zeta(s)$ *takes its values in* $\mathbb{C}\backslash[0,\infty)$ *for* $\Im m\, s > 0$.

**Remark.** If $e^{\zeta(s)} \in \hat{\mathscr{T}}$, then $\zeta(s)$ is a Pick function. It need not be if just $e^{\zeta(s)} \in \hat{\mathscr{T}_e}$ as the example $\zeta(s) = s^2/2$ shows. □

**Proof** (rather technical). Assume that $\zeta(s_0) = t_0 \geq 0$, $\Im m\, s_0 > 0$. Then

$$\frac{1}{2\pi i}\int \frac{\zeta'(s)}{\zeta(s)-t_0} ds = m, \qquad m \in \mathbb{N},$$

with the integral taken in the positive direction along a small circle around $s_0$. Approximating the given function $\zeta(s)$ uniformly on a compact domain by a function of the simpler type

$$\zeta(s) = \sum_{k=1}^{n} \log(\frac{t_k}{t_k - s})\, u_k + \sum_{k=1}^{n'} \log(\frac{t_k'}{t_k' + s})\, u_k' \quad \text{(positive parameters)}$$

so that moreover $\zeta'(s)$ is uniformly approximated (which can be done), we see that a contradiction is obtained if we can show that the lemma holds when $\zeta(s)$ has the simple form. Consider now a large contour with the appearance shown in Figure 7.1.1.

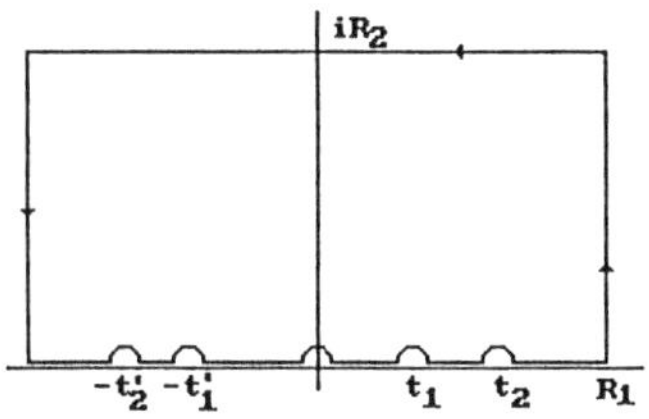


Figure 7.1.1

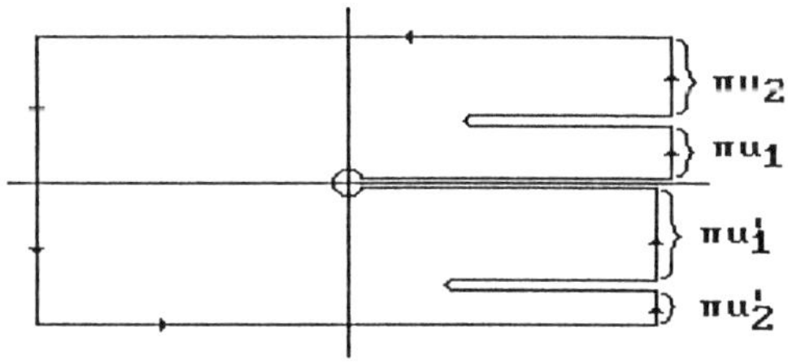


Figure 7.1.2

Here $R_1$ and $R_2$ are very large and the part of the contour along the real line is considered to be on its upper side. According to the principle of argument we now just have to show that when $s$ moves along this contour, the graph of $\zeta(s)$ does not cross $\mathbb{R}_+$. The appearance of the graph is shown in Figure 7.1.2.

In the figure the graph crosses the real line to the left of $0$, but that has to be verified. Consider $\Re e[\zeta(s)]$ for $-t_1' < s < t_1$. In this interval, $\Im m[\zeta(s)] = 0$. More precisely, for a small $\delta > 0$,

$$\Im m[\zeta(s+i\delta)] \approx \delta \cdot \left( \sum_{k=1}^{n} \frac{u_k}{t_k - s} - \sum_{k=1}^{n'} \frac{u_k'}{t_k' + s} \right).$$

Now

$$\Re e[\zeta(s)] = \sum_{k=1}^{n} \log\left|\frac{t_k}{t_k - s}\right| u_k + \sum_{k=1}^{n'} \log\left|\frac{t_k'}{t_k' + s}\right| u_k'$$

and hence, for $-t_1' < s < t_1$,

$$(\Re e[\zeta(s)])' = \sum_{k=1}^{n} \frac{u_k}{t_k - s} - \sum_{k=1}^{n'} \frac{u_k'}{t_k' + s} \qquad \left(= \lim_{\delta \to 0} \frac{\Im m[\zeta(s+i\delta)]}{\delta}\right)$$

which is negative for $s$ close to $-t_1'$ and positive for $s$ close to $t_1$. Further,

$$(\Re e[\zeta(s)])'' = \sum_{k=1}^{n} \frac{u_k}{(t_k-s)^2} + \sum_{k=1}^{n'} \frac{u_k'}{(t_k'+s)^2} > 0 .$$

Thus, for $-t_1' < s < t_1$, $\Re e[\zeta(s)]$ is a strictly convex function and has therefore a unique minimum at some point $s_{min}$. If $s_{min} \neq 0$, then as desired, $\Re e[\zeta(s_{min})] < 0$ since $\Re e[\zeta(0)] = 0$. If $s_{min} = 0$, this point is avoided by the small half–circle around 0, and the crossing takes again place as desired. □

**Remark.** If, as in Theorem 7.1.3, we assume that $\Im m[\zeta'(s)\overline{\zeta(s)}] \leq 0$ for $\Im m\, s > 0$, then it follows by a series expansion that $\zeta(s)$ takes its values in $\mathbb{C}\backslash[0,\infty)$ for $\Im m\, s > 0$. □

## 7.2 Examples

Some examples of genuine EGGC's are presented in this section. Further examples appear in Sections 7.3 and 7.4.

**Example 7.2.1** ***(Stable distribution).*** The general Stable distribution of index $\alpha$, $0 < \alpha \leq 2$, has chf of the form

$$\varphi(is) = \exp\{i\mu s - c\cdot|s|^{\alpha}\big(1 - i\beta\,\mathrm{sign}(s)\cdot\omega(s,\alpha)\big)\}, \quad s \in \mathbb{R},$$

where $\mu \in \mathbb{R}$, $c \geq 0$, $|\beta| \leq 1$, and $\omega(s,\alpha)$ equals $\tan(\alpha\pi/2)$ for $\alpha \neq 1$ and $-2\pi^{-1}\cdot\log|s|$ for $\alpha = 1$; see, e.g., Zolotarev (1986, p. 9). This expression, which has many variants, is hard to remember. It is symptomatic that it is often given with slight errors or at least lack of precision in many texts; cf. Hall (1981). The expression below for the mgf is simpler but seems to be little known; Zolotarev (1986, p. 112) gives it in a special case. The mgf $\varphi(s)$ is given by $e^{\zeta(s)}$ where

$$(7.2.1) \qquad \zeta(s) = \begin{cases} -c_1 s^{\alpha} - c_2(-s)^{\alpha} + \mu s & \text{if } 0 < \alpha < 1 \\ c_1 s^{\alpha} + c_2(-s)^{\alpha} + \mu s & \text{if } 1 < \alpha \leq 2 \\ c_1 s\log(s) + c_2(-s)\log(-s) + \mu s & \text{if } \alpha = 1 \end{cases}$$

with $c_1, c_2 \geq 0$ and $\mu \in \mathbb{R}$. Since $s^{\alpha} \sim s\cdot(1 + (\alpha-1)\log(s))$ as $\alpha \downarrow 1$, the expression for $\alpha = 1$ is obtained as a limit of that for $\alpha > 1$ with varying parameter values. Here $\zeta(s)$ is analytic in $\mathbb{C}\backslash\mathbb{R}$ and has different values on the two sides of the cut. It is easily verified that $e^{\zeta(is)}$, $s \in \mathbb{R}$, equals the chf of a Stable distribution; hence $e^{\zeta(s)}$ is the desired mgf.

For $0 < \alpha < 1$, $C_1 = 0$, $C_2 > 0$, and $\mu = 0$, a strictly Stable distribution on $\mathbb{R}_+$ is obtained. For $\mu = 0$ and $C_1 = C_2$, a symmetric Stable distribution appears. If, moreover, $\alpha = 1$, a Cauchy distribution with mgf

$$\varphi(s) = \begin{cases} \exp\{i\pi C\cdot s\} & \text{for } \Im m\, s > 0 \\ \exp\{-i\,\pi C\cdot s\} & \text{for } \Im m\, s < 0 \end{cases} \qquad (C = C_1 = C_2)$$

appears. To get this expression, one may use that $\arg[-s] = \arg[s] - \pi$ when $\arg[s] > 0$.

Now $\varphi'(s)/\varphi(s) = \zeta'(s)$ and

$$\zeta'(s) = \begin{cases} -C_1\alpha s^{\alpha-1} + C_2\alpha(-s)^{\alpha-1} + \mu & \text{if } 0 < \alpha < 1 \\ C_1\alpha s^{\alpha-1} - C_2\alpha(-s)^{\alpha-1} + \mu & \text{if } 1 < \alpha \leq 2 \\ C_1 \log(s) - C_2\log(-s) + C_1 - C_2 + \mu & \text{if } \alpha = 1 \end{cases}$$

Obviously $\zeta'(s)$ is a Pick function and hence the general Stable distribution belongs to $\mathcal{T}_e$. Further, for $\alpha < 2$, the U–measure has a density given by

$$u(t) = \frac{1}{\pi}\cdot\Im m[\zeta'(t+i\cdot 0)] = \begin{cases} C(t)\cdot\frac{\alpha|\sin(\alpha\pi)|}{\pi}\,|t|^{\alpha-1} & \text{if } \alpha \neq 1 \\ C(t) & \text{if } \alpha = 1 \end{cases}$$

where $C(t)$ equals $C_2$ for $t > 0$ and $C_1$ for $t < 0$. By (7.1.5), the corresponding Lévy density $\ell(y)$ is easily verified to be proportional to $|y|^{-1-\alpha}$ with different constants of proportionality for $y > 0$ and $y < 0$.

It is well known, see Feller (1971, p. 577), that if the iid rv's $X_j$, $j = 1, 2, ..., n$, have a pdf $f$ such that $f(x) \sim \tilde{C}_1\cdot x^{-1-\alpha}$ as $x \to \infty$ and $f(x) \sim \tilde{C}_2\cdot|x|^{-1-\alpha}$ as $x \to -\infty$, then the sum $\Sigma X_j$ can be normalized so as to be convergent in distribution to the general Stable distribution of index $\alpha$. Choosing $f$ to be the convolution of two different Pareto distributions, one on $(0, \infty)$ and the other on $(-\infty, 0)$, we see again by the closure theorem for EGGC's that the Stable distribution belongs to $\mathcal{T}_e$. □

**Example 7.2.2** ***(Maybe a surprise).*** Consider

$$\varphi(s) = \begin{cases} (1-is)^{-\gamma} & \text{for } \Im m\, s > 0 \\ (1+is)^{-\gamma} & \text{for } \Im m\, s < 0 \end{cases}$$

Now, for $\Im m\, s > 0$,

$$\Im m[\varphi'(s)/\varphi(s)] = \Im m[\frac{i\gamma}{1-is}] = \frac{\gamma\cdot(1+\Im m s)}{|1-is|^2} > 0.$$

Hence, since $\varphi(0) = 1$, $\varphi \in \hat{\mathscr{T}_e}$. The U–measure has a density given by

$$u(t) = \frac{1}{\pi}\cdot\Im m[\frac{\varphi'(t+i\cdot 0)}{\varphi(t+i\cdot 0)}] = \frac{\gamma}{\pi}\cdot(1+t^2)^{-1}, \quad t \in \mathbb{R}\setminus\{0\},$$

which is of the Cauchy type. The Lévy density admits no simple expression.

What is the corresponding distribution? The chf is given by $\varphi(is) = (1+|s|)^{-\gamma}$, and obviously it corresponds to a product $Y\cdot X$ of independent rv's such that $Y \sim \text{Gamma}(\gamma,1)$ and $X$ has a Cauchy distribution with chf $e^{-|s|}$. The result that $Y\cdot X \sim \mathscr{T}_e$ is also a consequence of the Composition Theorem. □

**Example 7.2.3** ***(Logarithm of Gamma variable).*** Let $X = \log Y$, where $Y \sim \text{Gamma}(\beta,1)$. The pdf and the mgf of $X$ are

$$f(x) = \frac{1}{\Gamma(\beta)}\exp\{\beta x - e^x\}, \; x \in \mathbb{R}, \qquad \varphi(s) = E[Y^s] = \frac{\Gamma(\beta+s)}{\Gamma(\beta)}.$$

Hence $\varphi'(s)/\varphi(s) = \psi(\beta+s)$, where here $\psi(z)$ is the derivative of $\log\Gamma(z)$. It is well known (Abramowitz & Stegun, 1964, p. 259) that

$$\psi(z) = -C + \sum_{k=1}^{\infty}\left(\frac{1}{k} + \frac{1}{-(k-1)-z}\right), \tag{7.2.2}$$

where $C$ is Euler's constant. Hence

$$\varphi'(s)/\varphi(s) = -C + \sum_{k=1}^{\infty}\left(\frac{1}{k} + \frac{1}{-(k-1+\beta)-s}\right), \tag{7.2.3}$$

which is obviously a Pick function; every term has positive imaginary part for $\Im m\, s > 0$. Thus $X \sim \mathscr{T}_e$. The U–measure has atoms of size 1 at the points $-\beta, -\beta-1, -\beta-2, \ldots$, which indicates that it should be possible to represent $X$ as a limit of essentially sums of independent negative Exponential rv's. In fact, by Euler's formula,

$$\Gamma(z) = \lim_{n\to\infty}\frac{n!n^z}{z(z+1)\cdot\ldots\cdot(z+n)},$$

or by an integration of (7.2.3), we easily obtain

$$\varphi(s) = \lim_{n\to\infty}\exp\{s\cdot\log n\}\prod_{j=0}^{n}\left(1+\frac{s}{\beta+j}\right)^{-1}$$

corresponding to, with $X_j \sim \text{Exp}(\beta+j)$,

$$X = \lim_{n\to\infty} \left(-\sum_{j=0}^{n} X_j + \log n\right) = -\sum_{j=0}^{\infty} \left(X_j - \frac{1}{\beta+j}\right) + \text{constant.}$$

Changing the location and scale of the distribution of $X$, we see that the logarithm of any power (positive or negative) of a general Gamma variable also has its distribution in $\mathcal{T}_e$. More generally, if $X$ is the logarithm of a product of powers of independent Gamma variables, then $X \sim \mathcal{T}_e$. □

**Example 7.2.4** ***(Generalized Logistic distribution).*** Consider $X = \log(Y_1/Y_2)$, with $Y_j \sim \text{Gamma}(\beta_j,1)$, $j = 1, 2$, independent. The pdf of $X$ is

$$f(x) = \frac{1}{B(\beta_1,\beta_2)} e^{-\beta_1 x}/(1+e^{-x})^{\beta_1+\beta_2}, \quad x \in \mathbb{R}.$$

This distribution, called the *generalized Logistic distribution,* obviously belongs to $\mathcal{T}_e$ and so it does when the location and scale are changed. In particular, the Logistic distribution obtained for $\beta_1 = \beta_2 = 1$ is an EGGC. In this case, the mgf is given by

$$\varphi(s) = \Gamma(1+s)\Gamma(1-s) = \frac{\pi s}{\sin(\pi s)} = \prod_{k=1}^{\infty} (1 - \frac{s^2}{k^2})^{-1} = \prod_{k=1}^{\infty} \frac{1}{(1-\frac{s}{k})(1+\frac{s}{k})}$$

showing that $X \stackrel{d}{=} \Sigma_{k=1}^{\infty}(X_k - \tilde{X}_k)$, where all the variables $X_k$ and $\tilde{X}_k$ are independent and $X_k, \tilde{X}_k \sim \text{Exp}(k)$. The pdf of $X_k - \tilde{X}_k$ is $k\cdot\exp\{-k|x|\}/2$; two–sided Exponential distribution.

For $\beta_1 = \beta_2 = 1/2$, we get

$$f(x) = \frac{1}{\pi}\cdot(e^{x/2}+e^{-x/2})^{-1} = \frac{1}{2\pi}\frac{1}{\cosh(x/2)}$$

and

$$\varphi(s) = \frac{1}{\pi}\cdot\Gamma(\tfrac{1}{2}+s)\Gamma(\tfrac{1}{2}-s) = \frac{1}{\cos(\pi s)} = \prod_{j=0}^{\infty} \left(1 - \frac{s^2}{(\frac{1}{2}+j)^2}\right)^{-1} = \prod_{j=0}^{\infty} \left(1 - \frac{s}{\frac{1}{2}+j}\right)^{-1}\left(1 + \frac{s}{\frac{1}{2}+j}\right)^{-1}.$$

Note that the chf $\varphi(ix)$ is proportional to $f(2\pi x)$; this property is called *self–reciprocity,* cf. Section 7.3. Convolving $f$ with itself, we get another EGGC with mgf $1/\cos^2(\pi s)$. By, e.g., the calculus of residues, the corresponding pdf can be shown to be, cf., Feller (1971, p. 64),

$$\frac{1}{\pi^2} x\cdot(e^{x/2} - e^{-x/2})^{-1} = \frac{1}{\pi^2}\frac{x/2}{\sinh(x/2)},$$

i.e., essentially the chf of a Logistic distribution. □

**Example 7.2.5** ***(Logarithm of positive Stable variable).*** Let X be a Stable variable with mgf $\varphi(s) = \exp\{-(-s)^\alpha\}$, $0 < \alpha \le 1$. We know from Example 5.6.2 that if $\alpha = 1/n$, $n = 2, 3, \ldots$, then $X^{-1}$ can be represented as a product of $n-1$ independent Gamma variables. Hence $\log X \sim \mathcal{T}_e$ in these cases and when $\alpha = 1$. What happens for other values of $\alpha$? We have, cf. Example 5.6.2,

$$\varphi(s) = \mathbf{E}[\exp\{s\log X\}] = \mathbf{E}[X^s] = \Gamma(1-\tfrac{s}{\alpha})/\Gamma(1-s).$$

Hence, by (7.2.2) and some simplification,

$$\frac{\varphi'(s)}{\varphi(s)} = -\frac{1}{\alpha}\psi(1-\tfrac{s}{\alpha}) + \psi(1-s) = C\cdot(\tfrac{1}{\alpha}-1) + \sum_{k=1}^{\infty}\Bigl(-\tfrac{1}{k}(\tfrac{1}{\alpha}-1) + \frac{1}{k\alpha-s} - \frac{1}{k-s}\Bigr).$$

It follows that

$$\Im m[\varphi'(s)/\varphi(s)] = \Im m\, s \sum_{k=1}^{\infty}\Bigl(\frac{1}{|k\alpha-s|^2} - \frac{1}{|k-s|^2}\Bigr). \tag{7.2.4}$$

For $\Im m\, s > 0$, this expression is negative when s is very close to 1 unless $\alpha = 1/n$ for some $n \ge 1$. Hence $\log X \sim \mathcal{T}_e$ iff $\alpha$ is the reciprocal of an integer. This negative result shows that the technique in Example 5.6.2 for proving that $X \sim \mathcal{B}$ for $\alpha = 1/n$ cannot be generalized. On the other hand, it is known that $\log X$ is self–decomposable for all $\alpha$ (Shanbhag, Pestana & Sreehari, 1977). □

In Examples 7.2.3 – 7.2.5 we have met distributions in the general PF∞–class, i.e. distributions which are limits of finite convolutions of Exponential distributions on $(-\infty,0)$ and $(0,\infty)$. The U–measure is discrete with atoms with integral mass. A distribution in this class can be represented as the convolution of a Normal distribution and a denumerable convolution of appropriately translated Exponential distributions on $(-\infty,0)$ and $(0,\infty)$. Note that if $X_t \sim \mathrm{Exp}(t)$ or $-X_t \sim \mathrm{Exp}(-t)$, $t < 0$, then $\mathrm{Var}[X_t - t^{-1}] = t^{-2}$ and hence the sum $\Sigma\,(X_t - t^{-1})$, taken over the atoms of the U–measure of an EGGC, is convergent almost surely. As mentioned in Chapter 3, the major document on PF∞–distributions is the book by Karlin (1968).

**Remark 1.** If $X \sim \mathrm{Exp}(1)$, then $e^X$ has the pdf $f(x) = x^{-2}$, $x > 1$, which equals the pdf of $Y_1/Y_2 + 1$, where $Y_1 \sim \mathrm{Exp}(1)$ and $Y_2 \sim \mathrm{Exp}(1)$ are independent. Hence any PF∞–distribution is the distribution of the logarithm of a product $P = e^Z\cdot\Pi_{j=1}^{\infty} a_j(Y_{1j}/Y_{2j} + 1)^{q_j}$, with $a_j > 0$, $q_j \in \mathbb{R}$, $Z \sim N(\cdot,\cdot)$, and $Y_{1j}$, $Y_{2j}$ as above. If $q_j \equiv 1$, then $P \sim \mathcal{T}$; cf. Theorem 6.2.2. □

**Remark 2.** It is not true that $X \sim PF\infty \Rightarrow e^X \sim \mathscr{B}$. In fact, $X = \log\sqrt{Y}$ with $Y \sim$ Gamma works as a counterexample. Nor is it true that $e^X \sim \mathscr{B} \Rightarrow X \sim PF\infty$. In fact, the pdf $C\cdot\exp\{-x-x^{-1}\}$, corresponding to $e^X$, belongs to $\mathscr{B}$, but the pdf of X is not even ID; cf. Example 5.6.3. □

### 7.3 EGGC's as Variance Mixtures of the Normal Distribution

Let $Z \sim N(0,1)$ and consider $X = \sqrt{Y}\cdot Z$, where $Y > 0$ is an rv independent of Z. The distribution of X is a *variance mixture* of the $N(0,\sigma^2)$–distribution; the mixing distribution equals that of Y. In this section we shall see that every symmetric EGGC is such a variance mixture with $Y \sim \mathscr{T}$. Many explicit pdf's in $\mathscr{T}_e$ are obtained by letting $Y \sim \mathscr{B}$. The corresponding class of distributions, $\mathscr{B}_e$, is closed with respect to multiplication and division of independent rv's. Student's t–distribution, in particular, belongs to $\mathscr{B}_e$.

The mgf of X above is given by $\psi(s) = \varphi(s^2/2)$, where $\varphi$ is the mgf of Y.

**Theorem 7.3.1.** *If* $\varphi$ *is the mgf of a GGC, then* $\psi(s) = \varphi(s^2/2)$ *is the mgf of a symmetric EGGC. Conversely, if* $\psi(s)$ *is the mgf of a symmetric EGGC, then* $\psi(s) = \varphi(s^2/2)$ *for some* $\varphi \in \hat{\mathscr{T}}$.

**Remark.** A chf $\chi(s)$, $s \in \mathbb{R}$, corresponds to a symmetric EGGC iff the function $\chi(\sqrt{x})$, $x > 0$, is HCM. This characterization follows from Theorems 7.3.1 and 6.1.1. □

The first part of the theorem easily follows from the Composition Theorem in Section 7.1. In fact, for $\zeta(s) = s^2/2$, we have, for $\Im m\, s > 0$, $\Im m[\zeta'(s)\overline{\zeta(s)}] = \frac{1}{2}|s|^2\Im m\,\bar{s} < 0$. However, it is illuminating to give a more direct proof to see how the two spectral measures are related to each other.

**Proof.** Let

$$\varphi(s) = \exp\{as + \int_{(0,\infty)} \log(\tfrac{t}{t-s})V(dt)\}, \qquad a \ge 0,\ V(dt) \ge 0.$$

Hence

$$\psi(s) = \exp\{as^2/2 + \int_{(0,\infty)} \Big(\log(\frac{\sqrt{2t}}{\sqrt{2t}+s}) + \log(\frac{\sqrt{2t}}{\sqrt{2t}-s})\Big)V(dt)\}$$

$$= \exp\Big\{as^2/2 + \int_{(-\infty,0)} \Big(\log(\tfrac{t}{t-s}) - \tfrac{st}{1+t^2}\Big)U(dt) + \int_{(0,\infty)} \Big(\log(\tfrac{t}{t-s}) - \tfrac{st}{1+t^2}\Big)U(dt)\Big\},$$

with U induced by V and the transformations $t \to -\sqrt{2t} = \tilde{t}$ and $t \to \sqrt{2t} = \tilde{t}$; $\tilde{t}$ is then renamed t. Since U is symmetric around 0, the two terms $st/(1+t^2)$, added to produce convergence, essentially cancel. The well–known conditions on the spectral measure (here called V) of a GGC imply that U satisfies the conditions (7.1.2). The first part of the theorem is proved; it is obvious that the distribution is symmetric (around 0).

We now turn to the converse. Let

$$\psi(s) = \exp\{bs + as^2/2 + \int \left(\log(\tfrac{t}{t-s}) - \tfrac{st}{1+t^2}\right)U(dt)\} \tag{7.3.1}$$

be the mgf of a symmetric EGGC. Then $\psi(s) = \psi(-s)$ and it follows that $\psi(s)$ can also be represented by (7.3.1) with b changed to $-b$ and U to $\tilde{U}$ with $\tilde{U}$ induced by U and the transformation $t \to -t$. The uniqueness of the representation (7.3.1) then guarantees that $b = 0$ and that $\tilde{U} = U$. Thus U is symmetric. By this symmetry, we then easily get

$$\psi(s) = \exp\left\{ as^2/2 + \int_{(0,\infty)} \log(\tfrac{t^2}{t^2-s^2})\ U(dt)\right\}.$$

The substitution $\tilde{t} = t^2/2$ gives the desired result. □

The well–known fact that any symmetric Stable variable X of index $\alpha$, $0 < \alpha < 2$, can be represented as $X \overset{d}{=} \sqrt{Y}\cdot Z$, with $Z \sim N(0,1)$ and independent of the positive and (strictly) Stable variable Y of index $\alpha/2$, illustrates Theorem 7.3.1. Another illustrating example is provided by the generalized Logistic distribution in Example 7.2.4 with $\beta_1 = \beta_2$. The U–measure for the variance mixture has only atoms of size 1. Of course, the same is true for the spectral measure of the mixing distribution. The variance mixture and the mixing distribution are both $PF_\infty$–distributions.

There is a slight extension of the first part of Theorem 7.3.1.

**Theorem 7.3.2.** *A $\sigma^2$–mixture of the $N(m\sigma^2,\sigma^2)$–distribution ($m \in \mathbb{R}$) belongs to $\mathcal{T}_e$ if the $\sigma^2$–mixing distribution is a GGC.*

**Proof.** Let $Y = \sigma^2$ have the mgf $\varphi$. The mgf of the mixture is given by $\psi(s) = \varphi(\zeta(s))$, where $\zeta(s) = ms + s^2/2$; $e^{\zeta(s)}$ is the mgf of the $N(m,1)$–distribution. The imaginary part of

$$\zeta'(s)/\zeta(s) = \frac{1}{s} + \frac{1/2}{m + s/2}$$

is negative for $\Im m\, s > 0$. By the Composition Theorem, the proof is complete. □

**Remark.** If $W(\cdot)$ is a Wiener–process, possibly with drift, and $T \geq 0$ is an independent random time and $T \sim \mathscr{T}$, then $W(T) \sim \mathscr{T}_e$; this is just a reformulation of Theorem 7.3.2. □

A simple consequence of Theorems 7.3.1 and 6.2.1 is the following result.

**Theorem 7.3.3.** *Let $X \sim \mathscr{T}_e$ and $Y \sim \mathscr{B}$ be independent rv's. If $X$ is symmetric, then $\sqrt{Y} \cdot X \sim \mathscr{T}_e$.*

In other words: A "variance mixture" of any symmetric EGGC with the mixing distribution in $\mathscr{B}$ is again a symmetric EGGC.

We now turn to the pdf's of variance mixtures of the $N(0,\sigma^2)$–distribution. Let $g_Y(y)$ be the pdf of $Y = \sigma^2 \sim \mathscr{T}$. The pdf corresponding to the mgf $\int_0^\infty \exp\{\frac{y}{2}s^2\}\, g_Y(y)dy$ is then given by

$$f(x) = (2\pi)^{-\frac{1}{2}} \int_0^\infty \exp\{-\frac{x^2}{2y}\}\, y^{-1/2} g_Y(y)dy \; = (2\pi)^{-\frac{1}{2}} \int_0^\infty \exp\{-\tfrac{y}{2}x^2\}\, \sqrt{y} g_{Y^{-1}}(y)dy \, .$$

We see that if also $C \cdot \sqrt{y} g_{Y^{-1}} \in \mathscr{T}$, then $\xi(s) = f(\sqrt{2s})$, $s > 0$, is proportional to the LT of a GGC. Obviously the case $Y \sim \mathscr{B}$ is of particular interest. By the properties of $\mathscr{B}$, we have that $g_Y \in \mathscr{B}$ iff $\sqrt{y} g_{Y^{-1}}$ is HCM. The first part of the following theorem is now obvious. The second part, less essential, is proved by the technique in the proof of Theorem 6.1.2.

**Theorem 7.3.4.** *Let $f$ be a symmetric pdf on $\mathbb{R}$. Then $f$ is a variance mixture of the $N(0,\sigma^2)$–distribution with the $\sigma^2$–distribution in $\mathscr{B}$ (or degenerate) iff $\xi(s) = f(\sqrt{2s})$, $s > 0$, is the LT of a HCM–function (or a degenerate function). Provided that $f(0+) > 0$, an equivalent condition is that, for all $n \in \mathbb{N}_0$, $(-1)^n \xi^{(n)}(s)$ is HCM.*

This theorem makes it possible to verify that some more or less well–known pdf's are EGGC's and hence ID and SD in particular. In view of the results in Section 5.7 on LT's of $\mathscr{B}$–densities, we find that a pdf of the form

$$f(x) = C \cdot \left(1 + c(x^2+\delta)^{1/k}\right)^{-\gamma} \quad (c > 0,\ \delta \geq 0,\ k = 1, 2, ...,\ \gamma > k/2)$$

is a symmetric EGGC. Of course, it is also an EGGC when the location is changed. Some well–known distributions appear for special or limit values of the parameters:

i) $f(x) = C\cdot(1+cx^2)^{-\gamma}$ ; *t–distribution, essentially*

ii) $f(x) = C\cdot(1+c|x|)^{-\gamma}$ ; *Two–sided Pareto distribution*

iii) $f(x) = C\cdot\exp\{-c\sqrt{x^2+\delta}\}$ ; *Hyperbolic distribution*

iv) $f(x) = C\cdot\exp\{-c|x|^{2/k}\}$ $(k = 1, 2, \ldots)$

**Remark.** Again, we see the desirability of $Y \sim \mathscr{B}$ when $Y$ is positive and Stable of index $\alpha \leq 1/2$. That result would imply that $f(x) = C\cdot\exp\{-c|x|^{\beta}\} \in \mathscr{T}_e$ for all $\beta \leq 1$. Incidentally, it is now obvious that for $\alpha > 1/2$ the Stable distribution is not in $\mathscr{B}$; the pdf f is not even ID for $1 < \beta < 2$ since $f(x)$ tends to 0 too rapidly as $x \to \pm\infty$. □

By Theorem 5.7.1, a more exotic symmetric pdf in $\mathscr{T}_e$ is given by

$$f(x) \;=\; C\cdot x^{-2}\cdot\frac{(1+cx^2)^{\epsilon}-1}{\epsilon}, \quad -1 \leq \epsilon < 1/2\,.$$

For $\epsilon = 0$, the interpretation is that

$$f(x) \;=\; C\cdot\frac{\log(1+cx^2)}{x^2} \qquad (C = \frac{\sqrt{c}}{2\pi}).$$

For $c = 1$ this latter distribution is somewhat more spread out than the Cauchy and may have future applications. By the calculus of residues its chf can be shown to be

$$\varphi(is) = e^{-s} - s\int_s^{\infty} y^{-1}e^{-y}dy\,, \quad s > 0.$$

The class of pdf's corresponding to $\sqrt{Y}\cdot Z$, with $Y \sim \mathscr{B}$ (or degenerate) and $Z \sim N(0,1)$ independent of Y, is called the *extended* $\mathscr{B}$*–class*, $\mathscr{B}_e$, although $\mathscr{B} \not\subset \mathscr{B}_e$. Of course, $\mathscr{B}_e \subset \mathscr{T}_e$. Moreover, $\mathscr{B}_e$ has some nice properties in common with $\mathscr{B}$.

**Theorem 7.3.5** ***(Multiplication theorem).*** *The class $\mathscr{B}_e$ is closed with respect to multiplication and division of independent rv's. Moreover, if $X_1 \sim \mathscr{B}_e$ and $X_2 \sim \mathscr{T}_e$, where $X_2$ is symmetric, are independent, then $X_1\cdot X_2 \sim \mathscr{T}_e$.*

**Proof.** Let $X_1 \sim \mathscr{B}_e$ and $X_2 \sim \mathscr{B}_e$ be independent rv's. They can be represented as $X_j = \sqrt{Y_j}\cdot Z_j$, $j = 1, 2$, where $Y_j \sim \mathscr{B}$ and $Z_j \sim N(0,1)$. All these rv's are independent. Now $X_1\cdot X_2 \overset{d}{=} (Y_1\cdot Y_2\cdot Z_2^2)^{\frac{1}{2}}\cdot Z_1$ and $X_1/X_2 \overset{d}{=} (Y_1\cdot Y_2^{-1}\cdot Z_2^{-2})^{\frac{1}{2}}\cdot Z_1$. Since $Z_2^2$ has a Gamma$(\frac{1}{2},\frac{1}{2})$–distribution, the first statement in the theorem is obvious from Theorem 5.1.1 (since $\mathscr{C} = \mathscr{B}$). If $X_2 \sim \mathscr{T}_e$ and is symmetric, then $Y_2 \sim \mathscr{T}$ and the second statement follows analogously from Theorems 5.1.1 and 6.2.1. □

Theorem 7.3.5 concerned multiplication of independent rv's. The next theorem concerns powers of rv's.

**Theorem 7.3.6.** *Let* $X \sim \mathcal{B}_e$. *Then* $|X|^q \sim \mathcal{B}$ *for all* $q \in \mathbb{R}$, $|q| \geq 2$. *Moreover,* $|X|^q sign(X) \sim \mathcal{B}_e$ *at least for all* $q \in \mathbb{N}$, $q \neq 2$, *but not always for* $q = 2$.

**Proof.** Let $X = \sqrt{Y} \cdot Z$, with $Y \sim \mathcal{B}$ and $Z \sim N(0,1)$. Since $Z^2 \sim$ Gamma, the first statement is obvious from the properties of the $\mathcal{B}$–class. To prove the first part of the second statement, it suffices to show that $|Z|^q sign(Z) \sim \mathcal{B}_e$ for $q \in \mathbb{N}$, $q \geq 3$. The pdf of this symmetric rv is $f(x) = C \cdot |x|^{\alpha-1} \exp\{-|x|^{2\alpha}/2\}$, where $\alpha = 1/q$. In view of Theorem 7.3.4, we must show that $\xi(s) = s^{-\gamma} \exp\{-s^{\alpha}\}$, with $\gamma = (1-\alpha)/2$, is the LT of an HCM–function. Obviously $\xi$ is proportional to the LT of $\int_0^x (x-y)^{\gamma-1} g(y)dy$, where $g$ is a Stable pdf of index $\alpha$ on $\mathbb{R}_+$. For at least $\alpha = 1/q$ ($q \in \mathbb{N}$, $q \geq 2$), we know that $g$ is HCM. Hence, by (II) in Section 5.8, $\xi$ is HCM if $y^{\gamma+1} g(y)$ is an increasing function. Now $y^{1+\alpha} g(y)$ has a finite positive limit as $y \to \infty$; see, e.g., Feller (1971, p. 583). Hence, by the properties of HCM–functions, $y^{\gamma+1} g(y)$ is indeed increasing for $\gamma \geq \alpha$, i.e., for $q = 1/\alpha \geq 3$. The desired result follows.

To verify the last part of the theorem, we show that the function $h(x) = \int_0^x (x-y)^{-3/4} y^{-3/2} \exp\{-y^{-1}/4\}dy$, corresponding to $q = 2$, is not HCM. By simple transformations we get

$$\tilde{h}(x) = x^{-3/4} h(1/x) = \int_0^\infty y^{-3/4} (y+x)^{1/4} \exp(-(y+x)/4\} \, dy .$$

If $h$ is HCM, then so is $\tilde{h}$. Since $\tilde{h}(0) > 0$, $\tilde{h}$ must then be decreasing. However, a differentiation shows that it is increasing in a neighbourhood of zero. This contradiction ends the proof. □

**Remark.** If $Z \sim N(0,1)$, then the pdf of $1/Z$ takes the value $0$ at $x = 0$ and hence $1/Z$ is not ID; Rohatgi, Steutel & Székely (1990). Thus it is not true that $X \sim \mathcal{B}_e \Rightarrow 1/X \sim \mathcal{B}_e$. Nor is it true that $f_1 \in \mathcal{B}_e$, $f_2 \in \mathcal{B}_e \Rightarrow C \cdot f_1 \cdot f_2 \in \mathcal{B}_e$. In fact, $f(x) = C \cdot \exp(-|x|^{2\alpha})$ is not ID for $1/2 < \alpha < 1$ but $f(x)$ is a limit of products of the form $C \cdot \Pi (1+c_j x^2)^{-\gamma_j}$. □

### 7.4 More on Composition

We saw in Section 7.3 that for a Wiener process $W(\cdot)$ and $T \sim \mathcal{T}$ necessarily $W(T) \sim \mathcal{T}_e$. In this section, which is not central, we consider a more general Stable (Lévy) process $X(\tau)$, $\tau \geq 0$, with mgf of the form $\exp\{\tau\zeta(s)\}$, where $\zeta(s)$ is given by (7.2.1). The mgf of $X(T)$ is $\varphi(\zeta(s))$, where $\varphi$ is the mgf of $T$.

**Theorem 7.4.1.** *Let $X(\tau)$, $\tau > 0$, be a Stable process of index $\alpha < 2$ with mgf $\exp\{\tau\zeta(s)\}$, where $\zeta(s)$ is defined by (7.2.1). Then, for $0 < \alpha < 1$, $X(T) \sim \mathcal{T}_e$ whenever $T \sim \mathcal{T}$ iff $\mu = 0$* **or** *$C_1 = 0$ and $\mu \geq 0$ (increasing $X(\cdot)$)* **or** *$C_2 = 0$ and $\mu \leq 0$ (decreasing negative $X(\cdot)$). For $1 < \alpha < 2$, the same holds iff $\mu = 0$* **or** *$C_1 = 0$ and $\mu \leq 0$* **or** *$C_2 = 0$ and $\mu \geq 0$. For $\alpha = 1$, the same holds iff $C_1 = C_2$ (Cauchy process, maybe with drift)* **or** *$C_1 = 0$ and $\mu \leq 0$* **or** *$C_2 = 0$ and $\mu \geq 0$.*

**Proof** (rather technical). In view of the Composition Theorem, we only have to verify that the inequality $\Im m[\zeta'(s)\overline{\zeta(s)}] \leq 0$ (or, equivalently, $\Im m[\zeta'(s)/\zeta(s)] \leq 0$ ) holds for $\Im m\, s > 0$ precisely when the above conditions are satisfied.

Consider first *the case* $1 < \alpha < 2$. By (7.2.1)

$$\zeta'(s) = C_1\alpha\cdot s^{\alpha-1} - C_2\alpha\cdot(-s)^{\alpha-1} + \mu = \frac{\alpha}{s}\cdot\zeta(s) + \mu\cdot(1-\alpha).$$

Hence

$$\zeta'(s)\overline{\zeta(s)} = \frac{\alpha}{s}\cdot|\zeta(s)|^2 + \mu\cdot(1-\alpha)\overline{\zeta(s)}\,. \tag{7.4.1}$$

If $\mu = 0$, then, obviously, $\Im m[\zeta'(s)\overline{\zeta(s)}] \leq 0$ for $\Im m\, s > 0$. For $s$ real and on the upper side of $\mathbb{R}$, the imaginary part of the first term on the right in (7.4.1) vanishes whereas

$$\mu\cdot(1-\alpha)\cdot\Im m[\overline{\zeta(s)}\,] = \begin{cases} -\mu(1-\alpha)C_1|s|^\alpha\sin(\alpha\pi)\,, & s < 0 \\ \mu(1-\alpha)C_2|s|^\alpha\sin(\alpha\pi)\,, & s > 0 \end{cases} \tag{7.4.2}$$

which is nonpositive iff $\mu = 0$ or $C_1 = 0$ and $\mu \leq 0$ or $C_2 = 0$ and $\mu \geq 0$. We must now verify that also in the last two cases $\Im m[\zeta'(s)/\zeta(s)] \leq 0$ for $\Im m\, s > 0$. For reasons of symmetry, it suffices to consider the first of these cases. Then $\zeta(s) = C_2(-s)^\alpha + a\cdot(-s)$, where $a = -\mu \geq 0$. Set $\psi(s) = -1/\zeta(s)$. Obviously,

$$\psi(s) = \frac{1}{s}\cdot(a + C_2(-s)^{\alpha-1})^{-1}.$$

The last factor is proportional to the mgf of a GGC. It easily follows that $\Im m[\psi'(s)/\psi(s)] \geq 0$ for $\Im m\, s > 0$; cf. the proof of Theorem 3.3.2. Since $\psi'/\psi = -\zeta'/\zeta$, we conclude that $\Im m[\zeta'(s)/\zeta(s)] \leq 0$ as desired for $\Im m\, s > 0$.

As to *the case* $\alpha < 1$, we easily find that (7.4.1) and (7.4.2) still hold. It follows that the given conditions on $C_1$, $C_2$, and $\mu$ are necessary. They are also sufficient. For instance, the sufficiency of the condition $C_1 = 0$ and $\mu \geq 0$ is a consequence of the Composition Theorem for GGC's (Theorem 3.3.1).

*Now let* $\alpha = 1$. In this case we have by (7.2.1)

$$\zeta'(s) = C_1 \log s - C_2 \log(-s) + C_1 - C_2 + \mu = \frac{1}{s}\log(\zeta(s)) + C_1 - C_2.$$

Hence

$$\Im m[\zeta'(s)\overline{\zeta(s)}] - |\zeta(s)|^2\, \Im m[\tfrac{1}{s}] + (C_1 - C_2)\Im m[\overline{\zeta(s)}\,],$$

which obviously is nonpositive for $\Im m\, s > 0$ if $C_1 = C_2$. For $s$ real and on the upper side of $\mathbb{R}$, we have

$$\Im m[\overline{\zeta(s)}\,] = \begin{cases} -C_1|s|\pi & \text{if } s < 0 \\ C_2|s|\pi & \text{if } s > 0 \end{cases}$$

The sign of $(C_1 - C_2)\Im m[\overline{\zeta(s)}\,]$ is changed at $s = 0$ if $C_1 \neq C_2$ and $C_1 \neq 0$ and $C_2 \neq 0$. It follows necessarily that $C_1 = C_2$ or $C_1 = 0$ or $C_2 = 0$. For brevity, looking only at the case $C_2 = 0$, we find

$$\frac{\zeta'(s)}{\zeta(s)} = \frac{1}{s}\cdot\left(1 + \frac{C_1}{C_1 \log s + \mu}\right).$$

If $\mu < 0$, then $\Im m[\zeta'(s)/\zeta(s)]$ is positive for some $s$ in $\Im m\, s > 0$ such that $-\log s$ is close to $\mu/C_1$. Hence $\mu \geq 0$ if $C_2 = 0$. In this case $\Im m[\zeta'(s)/\zeta(s)] \leq 0$ for $\Im m\, s > 0$. This follows from the corresponding result for $\alpha > 1$ by a reasoning with limits. □

Some partial generalizations of Theorem 7.4.1 are possible:

*i)* *If* $\zeta(s) = \Sigma\, c_k s^{\alpha_k}$ *(or* $\zeta(s) = \Sigma\, c_k(-s)^{\alpha_k}$*),* $1 \leq \alpha_k \leq 2$, $c_k > 0$, *then* $\varphi(\zeta(s)) \in \hat{\mathscr{T}_e}$ *whenever* $\varphi \in \hat{\mathscr{T}}$.

*ii)* *If* $\zeta(is) = -\Sigma\, d_k|s|^{\alpha_k}$ $(s \in \mathbb{R})$, $0 < \alpha_k \leq 2$, $d_k > 0$, *then* $\varphi(\zeta(s)) \in \hat{\mathscr{T}_e}$ *whenever* $\varphi \in \hat{\mathscr{T}}$; *the EGGC is symmetric. (Note that* $(is)^\alpha + (-is)^\alpha = 2|s|^\alpha \cos(\alpha\pi/2)$*).)*

In (i) $e^{\zeta(s)}$ is the mgf of a convolution of extreme Stable distributions of index $\alpha$, $1 < \alpha \leq 2$. We know already from Section 3.3 that when $\zeta(s) = -\Sigma\, c_k(-s)^{\alpha_k}$, $0 < \alpha_k \leq 1$, then $\varphi(\zeta(s)) \in \hat{\mathscr{T}}$. To prove (i), one may again use the factor technique in the proof of

Theorem 3.3.2.

In (ii), $e^{\zeta(is)}$ equals the chf of a convolution of symmetric Stable distributions. Now, for $s \in \mathbb{R}$, $\varphi(\zeta(is)) = \varphi(-\Sigma\, d_k(-(is)^2)^{\alpha_k/2})$. We know that $\varphi(-\Sigma\, d_k(-s)^{\alpha_k/2})$ is the mgf of a GGC and hence, by essentially Theorem 7.3.1, $\varphi(\zeta(is))$ is the chf of an EGGC. This gives the desired result.

## 7.5 Self–Reciprocal Densities

Let f be a symmetric pdf on $\mathbb{R}$. Its chf is here denoted $\hat{f}$. The pdf f is said to be *self–reciprocal* (SR) if

(7.5.1) $$\hat{f}(x) = \sqrt{2\pi}\, f(x)\ .$$

If somewhat more generally $\hat{f}(x) = C\cdot f(cx)$, $(C > 0,\ c > 0)$, then by a change of scale, we can arrange so that (7.5.1) holds; the constant $\sqrt{2\pi}$ is a consequence of Parseval's formula. The N(0,1)–density and $f(x) = (\sqrt{2\pi}\cosh(\sqrt{\pi/2}\ x))^{-1}$; cf. Example 7.2.4, are examples of SR pdf's. The class of such densities is surprisingly large as we shall see in this section. The classes $\mathscr{B}$ and $\mathscr{T}$ play a role, though not a dominant one, in the exposition below.

Let first f be a variance mixture of the $N(0,\sigma^2)$–density. The mixing density for $\sigma^2$ is denoted g. Now (7.5.1) reduces to

$$\int_0^\infty \exp\{-\tfrac{y}{2}x^2\}g(y)dy \;=\; \int_0^\infty \exp\{-\tfrac{x^2}{2y}\}\, y^{-1/2}g(y)dy\ .$$

The right hand side equals $\int_0^\infty \exp\{-\frac{y}{2}x^2\}\ y^{-3/2}g(y^{-1})dy$. Hence and by the uniqueness of the LT, f is SR iff

(7.5.2) $$g(y) = y^{-3/2}g(y^{-1})\ .$$

This shows that the class of SR densities is very large; (7.5.2) holds iff $y^{3/4}g(y)$ is symmetric with respect to the transformation $y \to 1/y$. The class $\mathscr{B}$ is particularly suitable to provide simple examples. Using the canonical representation of a pdf in $\mathscr{B}$, we easily find that the pdf of a variance mixture with $g \in \mathscr{B}$ is SR iff $g(y) = C\cdot y^{-3/4}h(y)h(y^{-1})$, where $h \in \check{\mathscr{T}}$. In particular

(i) $$g(y) = C\cdot y^{-3/4}\exp\{-c(y + y^{-1})\}$$

and

(ii) $$g(y) = C\cdot y^{-3/4}(1+y)^{-\gamma}(1+y^{-1})^{-\gamma} = C\cdot y^{\gamma-3/4}(1+y)^{-2\gamma}$$

lead to SR pdf's. For $\gamma = 3/4$ in (ii), the corresponding symmetric pdf f can be shown (by an integration by parts and a suitable substitution) to be given by

$$(7.5.3) \qquad f(x) = \frac{1}{\sqrt{2\pi}}\Big(1 - x\cdot\exp\{x^2/2\}\sqrt{2\pi}\cdot(1-\phi(x))\Big), \quad x > 0,$$

where $\phi$ is the cdf of the N(0,1)–distribution. Obviously, this symmetric pdf is an EGGC.

Every SR pdf cannot be represented as a variance mixture of the Normal distribution. However, there is a more general technique due to Teugels (1971) for obtaining all SR pdf's though not explicitly.

**Theorem 7.5.1.** *A symmetric pdf* $f$ *is SR* (*i.e.* $\hat{f} = \sqrt{2\pi}\cdot f$) *iff the* ***h–transform*** $h_f(s) = \int \exp\{-sx^2/2\} f(x)dx$, $s > 0$, (*i.e. the LT of the rv* $X^2/2$, $X \sim f$) *satisfies*

$$(7.5.4) \qquad h_f(s) = s^{-1/2} h_f(s^{-1})$$

**Proof.** We have

$$\int \exp\{-sy^2/2\}\, \hat{f}(y)dy = \int \exp\{-sy^2/2\} \left(\int e^{iyx} f(x)dx\right) dy$$

$$= \int \left(\int \exp\{-sy^2/2\} e^{iyx} dy\right) f(x)dx = \int \sqrt{2\pi}\, s^{-1/2} \exp\{-\tfrac{1}{s}x^2/2\} f(x)dx ,$$

i.e.,

$$(7.5.5) \qquad h_{\hat{f}}(s) = \sqrt{2\pi}\, s^{-1/2} h_f(s^{-1}).$$

Thus if $\hat{f}(x) = \sqrt{2\pi}\, f(x)$, then (7.5.4) holds. Conversely, if (7.5.4) holds, then, by (7.5.5), $h_{\hat{f}}(s) = \sqrt{2\pi}\, h_f(s)$. The uniqueness of the LT shows that $\hat{f}(x) = \sqrt{2\pi}\, f(x)$. □

The class $\mathscr{H} = \{h_f \,;\, \hat{f} = \sqrt{2\pi}\, f\}$ is closed with respect to limits and mixing. Moreover, by Theorem 7.5.1,

i) If $h \in \mathscr{H}$, then $h^{1/\alpha}(s^\alpha) \in \mathscr{H}$ for $0 < \alpha \le 1$

ii) If $h_1, h_2 \in \mathscr{H}$, then $h_1^{\alpha_1}\cdot h_2^{\alpha_2} \in \mathscr{H}$ for $\alpha_1 + \alpha_2 = 1$, $\alpha_1, \alpha_2 > 0$ .

The N(0,1)–density has the transform $h(s) = (1+s)^{-1/2}$. Applying, then, (i) twice, we find that (a) $h(s) = (1+\sqrt{s})^{-1}$ and (b) $h(s) = (1+\sqrt[4]{s})^{-2}$ also correspond to SR pdf's. Here h in (a) is the LT of $X^2/2 = Y^2\cdot Z$, where Y and Z are independent and $Y \sim \text{Exp}(1)$ and Z is positive and strictly Stable of index 1/2. In fact, the pdf of the symmetric rv X is given by (7.5.3). To see this, we note that if

$$f(x) = \int_0^\infty \exp\{-x^2/(2y)\}\, y^{-1/2} g(y)\, dy\ ,$$

then

$$h_f(s) = \int_0^\infty \frac{y^{-3/2} g(1/y)}{\sqrt{s+y}} dy\ .$$

For $g(y) = \frac{1}{2}(1+y)^{-3/2}$, yielding the pdf (7.5.3), some calculation gives $h_f(s) = (1+\sqrt{s})^{-1}$.

**Remark.** Restricting attention to symmetric variables X such that the distribution of $X^2/2$ is a GGC with left–extremity 0, we easily find from (7.5.4) that the pdf of X is SR iff the U–measure corresponding to $X^2/2$ is symmetric with respect to the mapping $t \to 1/t$ and has total $\int U(dt) = 1/2$. All pdf's that can be generated from $h(s) = (1+s)^{-1/2}$ (corresponding to the N(0,1)–distribution) by means of the operations (i) and (ii) above belong to this restricted class of SR pdf's. □

## 7.6 Complementary Results and Notes

I (***Riemann's hypothesis***). Let $\zeta(z) = \sum_{n=1}^\infty n^{-z}$ ($\Re z > 1$) here denote Riemann's $\zeta$–function, which is analytic in $\mathbb{C}$ except for a simple pole at $z = 1$ and has (trivial) zeros at $-2, -4, -6, \ldots$ Let further $\xi(z) = \frac{1}{2}z(z-1)\cdot\Gamma(\frac{1}{2}z)\cdot\pi^{-\frac{1}{2}z}\cdot\zeta(z)$ be Riemann's $\xi$–function; see, e.g., Edwards (1974, pp. 16–18). It is known that $\xi(z) = \xi(1-z)$ and hence $\xi(\frac{1}{2}+z) = \xi(\frac{1}{2}-z)$. In a letter dated February 1st, 1988, O. Thorin conjectured $\xi(\frac{1}{2})/\xi(\frac{1}{2}+is)$ to be the mgf of a symmetric EGGC and, equivalently, $\xi(\frac{1}{2})/\xi(\frac{1}{2}+\sqrt{s})$, $s > 0$, to be the LT of a GGC. In one respect his hypothesis is simple: It is not hard to show, as Thorin also did, that it is equivalent to *Riemann's famous hypothesis* stating that the only non–trivial zeros of the $\zeta$–function are on the line $\Re z = \frac{1}{2}$. In view of Theorem 6.1.1, Riemann's hypothesis can therefore be reformulated in a *real* way as follows:

**Conjecture:** *The function* $\xi(\frac{1}{2})/\xi(\frac{1}{2}+\sqrt{s})$, *s > 0, is HCM.*

**Notes.** Thorin (1978a) introduced the EGGC's, verified that they are self–decomposable, and proved Theorem 7.1.1. Theorem 7.1.2 appeared in B1981 whereas Theorem 7.1.3 (Composition Theorem) is new. Example 7.2.1 (Stable distribution) is due to Thorin. He did not give the explicit expression for the mgf of a Stable distribution. Several people have, independently of each other, been involved in Example 7.2.3. Maloshevskii (1972) might have been the first to show that the distribution of the logarithm of a Gamma variable is ID. Cf. Shanbhag & Sreehari (1977) and B1978. The generalized Logistic distribution in Example 7.2.4 was considered by Barndorff–Nielsen, Kent & Sorensen (1982) and later also by Lai & Moore (1984). Example 7.2.5 appeared in B1981.

Kelker (1971) studied ID of variance mixtures of the Normal distribution. In particular, he was interested in the possible ID of the t–distribution. Using properties of special functions, Grosswald (1976) proved its ID; cf. Section 4.5. Other proofs appeared later. Barndorff–Nielsen & Halgreen (1977) showed that the Hyperbolic distribution is ID. These particular results have been sources of inspiration for Sections 7.3 and 7.4.

Theorems 7.3.1 and 7.3.2 are in part due to Thorin (1978a). Theorem 7.3.3 is new whereas Theorem 7.3.4 with a long proof essentially appeared in B1981. Theorems 7.3.5 and 7.3.6 are new. They are in part inspired by results of Rohatgi, Steutel & Szekely (1990). Theorem 7.4.1 is new.

The results in Section 7.5 are essentially due to Teugels (1971).

# 8. GENERALIZED NEGATIVE BINOMIAL CONVOLUTIONS

In Chapter 7 the class $\mathcal{T}$ was extended to contain distributions on $\mathbb{R}$ also. In this chapter we consider instead a class of probability distributions on $\mathbb{N}_0 = \{0, 1, 2, ...\}$, this being the discrete analogue of $\mathcal{T}$. The analogue of the Gamma distribution is the Negative Binomial distribution. By a *generalized Negative Binomial convolution (GNBC)*, we mean a limit distribution for a sequence of finite convolutions of Negative Binomial distributions. The class of GNBC's is denoted $\mathcal{T}_d$. Many of the results valid for $\mathcal{T}$ have their counterparts for $\mathcal{T}_d$, but there are also several complications.

In Section 8.1 the GNBC's are introduced and several of their properties are listed. Examples are presented in Section 8.2. GNBC's with explicit probability functions are the focus of Section 8.3. There is no known discrete analogue of the HCM–condition.

## 8.1 Definition and Main Properties

The Negative Binomial distribution, denoted $NB(\beta,q)$, has probability function (pf)

$$p_j = (-1)^j\binom{-\beta}{j}q^jp^\beta, \quad j \in \mathbb{N}_0 \qquad (0 \le q < 1,\ p = 1-q,\ \beta > 0)$$

and probability generating function (pgf)

$$P(z) = \Sigma_{j=0}^{\infty} p_jz^j = \left(\frac{p}{1-qz}\right)^\beta = \left(1 - \frac{q}{p}(z-1)\right)^{-\beta}.$$

The coefficient $(-1)^j\binom{-\beta}{j}$ can also be expressed as $\binom{\beta+j-1}{j}$ or as

$$\frac{\Gamma(\beta+j)}{\Gamma(\beta)\,j!} = \frac{\beta\cdot(\beta+1)\cdot\ldots\cdot(\beta+j-1)}{j!} = \frac{\beta^{[j]}}{j!} \qquad (\beta^{[0]} = 1)\,.$$

Obviously $P(z) = \varphi(z-1)$, where $\varphi$ is the mgf of a Gamma($\beta$, p/q)–distribution; the NB($\beta$,q)–distribution is a Po($\lambda$)–mixture (Po = Poisson) with $\lambda \sim \text{Gamma}(\beta, t)$, $t = p/q$. Looking directly at the pf we have, with $c = t^{-1} = q/p$ and $\lambda = c\tilde{\lambda}$,

$$(8.1.1)\quad \mathbf{E}[\tfrac{\lambda^j}{j!}e^{-\lambda}] = \tfrac{c^j}{j!}\,\mathbf{E}[(\tilde{\lambda})^je^{-c\tilde{\lambda}}] = \tfrac{c^j}{j!}(-1)^j\,\tfrac{d^j}{dc^j}(1+c)^{-\beta} = (-1)^j\binom{-\beta}{j}\left(\tfrac{c}{1+c}\right)^j\left(\tfrac{1}{1+c}\right)^\beta = p_j\,.$$

We define a GNBC to be a (non–defective) weak limit of a sequence of finite convolutions of Negative Binomial distributions and denote the class of GNBC's by $\mathcal{T}_d$. Since convolution corresponds to multiplication of pgf's or mgf's and a weak limit corresponds to a pointwise limit of pgf's (in the region $-1 \leq z \leq 1$) or mgf's (in e.g. the region $-2 \leq s \leq 0$), it follows that a distribution on $\mathbb{N}_0$ is a GNBC iff its pgf $P(z)$ equals $\varphi(z-1)$ with $\varphi \in \hat{\mathcal{T}}$ (i.e., $\varphi$ is the mgf of a GGC). Equivalently, a distribution belongs to $\mathcal{T}_d$ iff it is a Po($\lambda$)–mixture with $\lambda \sim \mathcal{T}$.

***Canonical Representation.*** *The pgf of a GNBC has the canonical representation*

$$(8.1.2) \qquad P(z) = exp\{a(z-1) + \int_{(0,1)} log(\tfrac{p}{1-qz}) V(dq)\},$$

*where* $a \geq 0$ *and* $V(dq)$ *is a nonnegative measure on* $(0,1)$ *satisfying* $\int_{(0,1/2]} qV(dq) < \infty$ *and* $\int_{(1/2,1)} log(p) V(dq) < \infty$.

This representation is an immediate consequence of the corresponding representation of the mgf of a GGC. The measure $V(dq)$ is induced by $U(dt)$ and the transformation $t \to (t+1)^{-1} = q$. We have $a = \lim_{z\to-\infty} \frac{\log(P(z))}{z-1}$. If the V–measure has a density, then, in view of e.g. (3.1.7), it is given by

$$(8.1.3) \qquad v(q) = \frac{1}{\pi}\cdot\frac{1}{q^2} \times \lim_{z\to1/q;\ \Im m\, z > 0} \text{es}\ \Im m[P'(z)/P(z)] .$$

We have $v(q) = u(t)\cdot\frac{dt}{dq} = q^{-2}u(t)$. If $a = 0$, then the total V–measure is given by

$$\int_{(0,1)} V(dq) = \lim_{z\to-\infty} \left(-zP'(z)/P(z)\right).$$

Obviously, $\mathcal{T}_d$ is closed with respect to convolution and weak limits. Many other results for $\mathcal{T}_d$ follow in some way or another from the 1–1 correspondence between distributions in $\mathcal{T}_d$ and $\mathcal{T}$. Some properties and characterizations are given below for the GNBC's. The first property is rather obvious.

***i)*** ***Thinning.*** *Let* $\epsilon_1, \epsilon_2, \epsilon_3, \ldots$ *be independent Bin(1,$\pi$)–variables,* $0 \leq \pi \leq 1$, *and independent of* $X \sim \mathcal{T}_d$. *Then* $\Sigma_{j=1}^{X} \epsilon_j \sim \mathcal{T}_d$. *Its pgf is* $P(1-\pi + \pi z)$, *where* $P$ *corresponds to* $X$.

***ii)*** ***Approximation Property.*** *If* $\lambda \sim \mathcal{T}$, *then there are rv's* $J_n \sim \mathcal{T}_d$, $n = 1, 2, \ldots$, *such that* $J_n/n \to \lambda$ *in distribution as* $n \to \infty$. *Conversely, if* $J_n \sim \mathcal{T}_d$, $n = 1, 2, \ldots$, *and* $J_n/n \to \lambda$ *in distribution, then* $\lambda \sim \mathcal{T}$.

**Proof.** Let $J_n \sim \mathcal{T}_d$ have pgf $\varphi_{n\lambda}(z-1)$; $n\lambda \sim \mathcal{T}$. Then, for $s < 0$,

$$\varphi_{J_n/n}(s) = \mathbf{E}[\exp\{sJ_n/n\}] = \mathbf{E}[(e^{s/n})^{J_n}] = \varphi_{n\lambda}(e^{s/n}-1) = \varphi_\lambda(n(e^{s/n}-1)) \to \varphi_\lambda(s).$$

which gives the first part of (ii). As to the converse, we have

$$P_{J_n}(e^{s/n}) = \mathbf{E}[\exp\{sJ_n/n\}] \to \varphi_\lambda(s), \quad s < 0.$$

Since $P_{J_n}(e^{s/n}) = \psi_n(n(e^{s/n}-1))$ with $\psi_n \in \hat{\mathcal{T}}$ and $n(e^{s/n}-1) \to s$, it easily follows that $\psi_n(s) \to \varphi_\lambda(s)$. By the closure theorem for $\mathcal{T}$, $\lambda \sim \mathcal{T}$. □

***iii) Mixtures.*** *A GNBC with $a = 0$ (i.e. without Poissonian convolution factor) and $\int V(dq) = \beta$, $0 < \beta < \infty$, is a $q$–mixture of the $NB(\beta, q)$ distribution. Moreover, any such $q$–mixture is a GNBC if the (random) odds $W = q/p \sim \mathcal{T}$.*

**Proof.** Let $P$ be the pgf of the GNBC. Let $\lambda \sim \mathcal{T}$ have mgf $\varphi(s) = P(s+1)$. Then, according to Theorem 4.1.1, $\lambda \overset{d}{=} \tilde{\lambda}\cdot W$, where $\tilde{\lambda} \sim \text{Gamma}(\beta,1)$ and $W > 0$ are independent. Now

$$\text{(8.1.4)} \qquad p_j = \mathbf{E}[\tfrac{\lambda^j}{j!}e^{-\lambda}] = \mathbf{E}[(-1)^j\tbinom{-\beta}{j}(\tfrac{W}{1+W})^j(\tfrac{1}{1+W})^\beta],$$

cf. (8.1.1). Thus the distribution of $q$ equals the distribution of $W/(1+W)$.

The second (and deeper) statement in (iii), is a consequence of Theorem 6.2.1 with $Y = \tilde{\lambda}$ and $X = W = q/p$. By this theorem $\tilde{\lambda}\cdot W \sim \mathcal{T}$. □

***iv) Power Series Distribution.*** *If the pf $\{p_j\}_0^\infty$ belongs to $\mathcal{T}_d$, then so does the pf $\tilde{p}_j = \frac{1}{P(\theta)}p_j\theta^j$, $j \in \mathbb{N}_0$, $0 < \theta \le 1$.*

**Proof.** Let $P(z) = \varphi_\lambda(z-1)$; $\lambda \sim \mathcal{T}$. Then

$$\tilde{P}(z) = \frac{1}{P(\theta)}P(\theta z) = \frac{1}{P(\theta)}\varphi_\lambda(\theta z-1) = \frac{1}{P(\theta)}\,\varphi_\lambda\big(\theta(z-1)-(1-\theta)\big) = \psi(z-1).$$

Clearly $\psi(s) = \frac{1}{P(\theta)}\varphi_\lambda(\theta s-(1-\theta)) \in \hat{\mathcal{T}}$; the pdf of $\theta\lambda$ is multiplied by an exponential factor and then normalized. □

***v) Complex Characterization.*** *A pf on $\mathbb{N}_0$ is a GNBC iff its pgf $P$ is analytic in $\mathbb{C}\backslash[1,\infty)$ and **(a)** $\Im m[P'(z)\overline{P(z)}] \ge 0$ in $\Im m\, z > 0$ and **(b)** $\lim_{z\downarrow-\infty} P'(z)/P(z) \ge 0$*

With $\varphi(s) = P(s+1)$, this result is a consequence of the proofs of Theorems 3.1.2 and 3.1.3. Condition (**b**) cannot be deleted; counterexample: $P(z) = \exp\{a(z^2-1)\}$, $a > 0$.

*vi)* ***Real Characterization.*** *A pf* $\{p_j\}_0^\infty$ *is a GNBC iff the sequence* $\{r_k\}_0^\infty$ *defined by*

$$(j+1)p_{j+1} = \Sigma_{k=0}^{j} p_{j-k}r_k \tag{8.1.5}$$

*is a completely monotone (CM) sequence.*

**Proof.** Obviously (8.1.5) is equivalent to $P'(z) = P(z)R(z)$. For a GNBC we have

$$\frac{P'(z)}{P(z)} = a + \int_{(0,1)} \frac{qV(dq)}{1-qz}$$

and hence

$$r_0 = a + \int_{(0,1)} qV(dq), \qquad r_k = \int_{(0,1)} q^k qV(dq), \quad k \geq 1, \tag{8.1.6}$$

which shows that $\{r_k\}_0^\infty$ is CM. Conversely, if $\{r_k\}_0^\infty$ is CM, then, by Hausdorff's characterization,

$$r_k = \int_{[0,1]} q^k H(dq), \quad k \in \mathbb{N}_0, \quad H(dq) \geq 0.$$

Since $\Sigma\, p_j = 1$, it easily follows that $H(\{1\}) = 0$. Setting $a = H(\{0\})$ and $V(dq) = q^{-1}H(dq)$ for $q > 0$, we get (8.1.5). The rest is obvious. □

For a particular pf, computer results can make it probable that the CM–condition in (v) holds or prove that it does not. It also follows from (v) that a GNBC is discrete self–decomposable and hence unimodal; cf. Section 2.3. A reformulation of (vi) states that a pgf $P$ corresponds to a GNBC iff, for all $n \in \mathbb{N}_0$, $\left((z-1)^n \frac{P'(z)}{P(z)}\right)^{(n)}$ is absolutely monotone on $(0,1)$, cf. Theorem 3.4.1.

*vii)* ***Curious Mixing.*** *If P is the pgf of a GNBC and* $g \in \mathscr{B}$, *then* $\int_0^\infty P(1-y+zy)g(y)dy$ *is the pgf of a GNBC. Further, for any* $g \in \mathscr{T}$, $\int_0^\infty P(z-y)g(y)dy$ *is proportional to the pgf of a GNBC.*

The results in (vii) are essentially just reformulations of Theorem 6.2.1 and the corollary of Theorem 6.2.4.

We end this section with a particular relation between GNBC's and GGC's.

**Theorem 8.1.1.** *Let* $Y_j$, $j \in \mathbb{N}$, *and* $X$ *be mutually independent rv's such that* $Y_j \sim Exp(1)$ *and* $X \sim \mathscr{T}_d$. *Let the V–measure corresponding to* $X$ *have total mass* $\beta$ *and let* $Y_0 \sim Gamma(\beta,1)$ *be an rv independent of all the others. Then* $X_1 = \Sigma_{j=0}^{X} Y_j \sim \mathscr{T}$. *Moreover, if* $Z_0 \sim Gamma(2\beta,1)$, *then* $X_2 = Z_0 + \Sigma_{j=1}^{2X} Y_j \sim \mathscr{T}$.

**Proof.** Let P be the pgf of X and $\varphi_i$ the mgf of $X_i$, i = 1, 2. Obviously,

$$\varphi_1(s) = P(\tfrac{1}{1-s})\cdot(\tfrac{1}{1-s})^{\beta} \qquad \text{and} \qquad \varphi_2(s) = P((\tfrac{1}{1-s})^2)\cdot(\tfrac{1}{1-s})^{2\beta}.$$

For $\varphi_1$, after some calculation, we easily get

$$\varphi_1'(s)/\varphi_1(s) = \int_{(0,1)} \left(\frac{1}{p-s} - \frac{1}{1-s}\right) V(dq) + \frac{\beta}{1-s} = \int_{(0,1)} \frac{1}{p-s} V(dq).$$

Since this is a Pick–function, the first result follows; the U–measure has total $\beta$ and all its mass in (0, 1). For $\varphi_2$, after much calculation, we get

$$\frac{\varphi_2'(s)}{\varphi_2(s)} = \int_{(0,1)} \left(\frac{2(1-s)}{(1-s)^2-q} - \frac{2}{1-s}\right) V(dq) + \frac{2\beta}{1-s} = \int_{(0,1)} \left(\frac{1}{1+\sqrt{q}-s} + \frac{1}{1-\sqrt{q}-s}\right) V(dq)$$

Since the last expression is a Pick–function, the second result follows. The total U–measure corresponding to $X_2$ is $2\beta$. □

## 8.2 Examples of Distributions in $\mathscr{T}_d$

Distributions which are limits of finite convolutions of geometric distributions are obviously in $\mathscr{T}_d$ and form a subclass of the class of discrete $PF_\infty$–distributions. The general $PF_\infty$–distribution on $\mathbb{N}_0$ has an additional convolution factor that is a denumerable convolution of Binomial distributions; cf. Karlin (1968, Chapter 8). We now give some other more explicit examples of GNBC's.

**Example 8.2.1** ***(Discrete Stable distribution).*** A pf with pgf of the form $P(z) = \exp\{-c(1-z)^\alpha\}$, $c > 0$, $0 < \alpha \le 1$, is called discrete (strictly) Stable; cf. Steutel & van Harn (1979). Since $\varphi(s) = P(s+1) = \exp\{-(-s)^\alpha\}$ belongs to $\hat{\mathscr{T}}$, the pf is a GNBC. It is discrete Stable in the sense that if $X_1$ and $X_2$ are independent rv's with the pgf's $P_j(z) = \exp\{-c_j(1-z)^\alpha\}$, j = 1, 2, then the pgf of the rv

$$X = \Sigma_{i=1}^{X_1} \epsilon_{1i} + \Sigma_{i=1}^{X_2} \epsilon_{2i},$$

where $\epsilon_{ji} \sim \text{Bin}(1,\pi_j)$ are independent rv's, is given by $P(z) = \exp\{-c(1-z)^\alpha\}$ with $c = c_1(1-\pi_1)^\alpha + c_2(1-\pi_2)^\alpha$. □

**Example 8.2.2** ***(First passage time).*** Consider a random walk in discrete time on $\mathbb{Z}$ with start at the point 0. At each of the times 1, 2, 3, ... a step is taken to the right with probability $p^o$ or to the left with probability $q^o = 1-p^o$. The time $T_1$ to reach the point 1 is of interest. As this point can only be reached at the times 1, 3, 5, ... , we consider instead $T = (T_1-1)/2$. The pgf of $T_1$ satisifes the equation

$$P_{T_1}(z) = p^o z + q^o z (P_{T_1}(z))^2.$$

Solving this quadratic equation and using that the pgf of $T$ is given by $P(z) = P_{T_1}(\sqrt{z})/\sqrt{z}$, we find that

$$P(z) = \frac{1}{2q^o}\frac{1-\sqrt{1-\kappa z}}{z} = \frac{\kappa}{2q^o}\frac{1}{1+\sqrt{1-\kappa z}}; \qquad \kappa = 4p^o q^o.$$

Here $P(0) = p^o$ and $P(1) = 1$ iff $p^o \geq 1/2$ which we assume holds. Certainly $P$ is analytic in $\mathbb{C}\setminus[1,\infty)$ and zero–free. By differentiation,

$$\frac{P'(z)}{P(z)} = \frac{\kappa/2}{\sqrt{1-\kappa z}\cdot(1+\sqrt{1-\kappa z})} = \frac{1}{\sqrt{1-\kappa z}}\cdot\frac{1}{2z} - \frac{1}{2z}.$$

Now it is not hard to verify that $\Im m[P'(z)/P(z)] \geq 0$ for $\Im m\, z > 0$ and since, moreover, $P'(z)/P(z) \to 0$ as $z \to -\infty$, it follows by (v) in Section 8.1 that $T \sim \mathcal{G}_d$. The density of the spectral measure $V$ is found by (8.1.3) and equals

$$v(q) = \frac{1}{2\pi}\cdot q^{-1/2}(\kappa-q)^{-1/2}, \quad 0 < q < \kappa,$$

i.e. a Beta(1/2, 1/2) density on $(0,\kappa)$ with, perhaps surprisingly, total mass 1/2; cf. Example 3.2.3 (random walk in continuous time).

The first passage time in Example 3.2.3, now called $T'$, can be expressed as $T' = \Sigma_{j=0}^{2T} Y_j$, where the $Y_j$'s are independent Exp(1)–variables. In view of the second part of Theorem 8.1.1, it follows again that $T' \sim \mathcal{G}$ and that the corresponding U–measure has total mass 1. □

**Example 8.2.3** ***(Generalized Waring distribution).*** Consider

$$\text{(8.2.1)} \qquad p_j = C\cdot\frac{\beta^{[j]}\cdot\alpha^{[j]}}{j!(\alpha+\beta+\gamma)^{[j]}}, \quad j \in \mathbb{N}_0,$$

where $\alpha$, $\beta$, and $\gamma$ are positive and $\beta^{[j]}$ is defined in Section 8.1. This distribution is known as the *generalized Waring distribution*; see, e.g., Irwin (1975). For $\beta = 1$ the

*Waring distribution* appears; cf. Johnson & Kotz (1969, p. 250). If moreover $\alpha = 1$ and $\gamma \to 0$, the *Yule distribution* appears. These distributions have been used for modelling word size in prose, for instance. The NB($\beta$,q)–distribution is obtained if we set $\alpha/(\alpha+\gamma) = q$ and let $\alpha$ and $\gamma$ tend to $\infty$.

Consider now a Po($\lambda$)–mixture with $\lambda = Y \cdot X_1/X_2$, where $Y$, $X_1$, and $X_2$ are independent rv's and $Y \sim \text{Gamma}(\beta,1)$, $X_1 \sim \text{Gamma}(\alpha,1)$, and $X_2 \sim \text{Gamma}(\gamma,1)$. By Theorem 5.1.1, $\lambda \sim \mathscr{B} \subset \mathscr{T}$ and hence the mixture is a GNBC. The corresponding pf is given by $p_j = (-1)^j\binom{-\beta}{j} E[q^j p^\beta]$, where $q = \frac{X_1/X_2}{1 + X_1/X_2} = X_1/(X_1+X_2)$; cf. (8.1.4). Here $(-1)^j\binom{-\beta}{j} = \beta^{[j]}/j!$. Further, since $q \sim \text{Beta}(\alpha, \gamma)$, we get

$$E[q^j p^\beta] = \int_0^1 q^j(1-q)^\beta \frac{1}{B(\alpha,\beta)} q^{\alpha-1}(1-q)^{\gamma-1}dq = \frac{B(j+\alpha,\beta+\gamma)}{B(\alpha,\beta)} = C \cdot \alpha^{[j]}/(\alpha+\beta+\gamma)^{[j]},$$

where $C = \frac{\Gamma(\beta+\gamma)\,\Gamma(\alpha+\gamma)}{\Gamma(\alpha+\beta+\gamma)\Gamma(\gamma)}$. Hence the Po($\lambda$)–mixture equals the generalized Waring distribution which thus is a GNBC.

The pgf P is the sum of a Hypergeometric series. In the special case $\beta = 1$, $\alpha = 1-\epsilon$, $\gamma = \epsilon$ $(0 < \epsilon < 1)$, we have

$$p_j = \epsilon \cdot \frac{(1-\epsilon)^{[j]}}{(j+1)!}, \quad j \in \mathbb{N}_0, \quad \text{and} \quad P(z) = \frac{(1-z)^\epsilon - 1}{-z}$$

as is easily verified by a series expansion. It follows that $\phi(s) = P(1-s) = (s^\epsilon-1)/(s-1)$ is the LT of a GGC (with pdf in $\mathscr{B}$.) In fact, this was already found in Section 5.7. □

**Example 8.2.4** ***(Logarithmic series distribution).*** Multiplying $p_j$ in (8.2.1) by $\theta^j$ $(0 < \theta < 1)$ and then setting $\beta = \alpha = 1$ and letting $\gamma \to 0$, we get, by (iv) in Section 8.1, a GNBC with pf

$$p_j = C \cdot \frac{1}{j+1} \theta^j, \quad j \in \mathbb{N}_0.$$

This is the *Logaritmic series distribution* on $\mathbb{N}_0$, cf. Section 2.3. Its pgf is given by

$$P(z) = \frac{1}{\log(1-\theta)} \cdot \frac{\log(1-\theta z)}{z}.$$

It follows that $P(1-s)$ and $P(-s)/P(0) = \frac{\log(1+\theta s)}{\theta s}$ are LT's of pdf's in $\mathscr{T}$ (and $\mathscr{B}$); cf. Section 5.7. □

### 8.3 In Search of a Discrete $\mathscr{B}$–class

When the mgf $\varphi$ runs through $\hat{\mathscr{T}}$, then $p_j = \varphi^{(j)}(-1)/j!$, $j \in \mathbb{N}_0$, describes all pf's in $\mathscr{T}_d$. This is seen from the McLaurin expansion of $\varphi(z-1)$. However, it is desirable to have large subclasses of more explicit pf's in $\mathscr{T}_d$. As is now well known, a pdf of the form $f(x) = C\cdot x^{\beta-1}h(x)$, $x > 0$, where $\beta > 0$ and $h$ is the LT of a GGC, belongs to $\mathscr{T}$. The discrete analogous result would be that pf's of the form

$$p_j = C\cdot(-1)^j\binom{-\beta}{j}\prod_{i=1}^{n}(t_i+j)^{-\gamma_i}, \quad j \in \mathbb{N}_0,$$

(all parameters positive) and limits thereof belong to $\mathscr{T}_d$. However, without some restriction on the parameters this result does not hold. For instance, the pf $p_j = C\cdot\exp\{-j^\alpha\}$, $j \in \mathbb{N}_0$, does not belong to $\mathscr{T}_d$ if $\alpha \lesssim 0.46$. In fact, as is easily verified numerically, if $\alpha \lesssim 0.46$, then $r_1 > r_0$; cf. (vi) in Section 8.1. On the other hand, numerical calculations indicate that for the pf $p_j = C\cdot\exp\{-(j+1)^\alpha\}$, $j \in \mathbb{N}_0$, $(0 < \alpha < 1)$, the sequence $\{r_k\}_0^\infty$ is CM and hence that this pf is a GNBC.

We now consider more generally a q–mixture of the NB($\beta$,q) distribution and represent it as

$$p_j = (-1)^j\binom{-\beta}{j}E[q^jp^\beta] = (-1)^j\binom{-\beta}{j}\int_0^\infty e^{-jx}k(x)dx\,. \tag{8.3.1}$$

Here $k(x) = e^{-x}(1-e^{-x})^\beta f_q(e^{-x})$ is obtained by the substitution $q = e^{-x}$ in the integral giving $E[q^jp^\beta]$. By (iii) in Section 8.1 we know that any GNBC, without Poissonian factor and with $\int V(dq) = \beta$, has this form. Moreover, if $W = q/p \sim \mathscr{T}$, then $\{p_j\}_0^\infty$ in (8.3.1) is a GNBC. We have

$$f_W(y) = f_q\left(\frac{y}{1+y}\right)\frac{1}{(1+y)^2} = \frac{(1+y)^{\beta-1}}{y}k\left(\log(1+\frac{1}{y})\right).$$

Hence a *sufficient condition* on $k$ for (8.3.1) to yield a GNBC is that the last expression above is the pdf of a GGC. There is also a more important *necessary and sufficient condition:*

**Theorem 8.3.1.** *A pf of the form (8.3.1) is a GNBC iff*

$$g(t) = \frac{t^{\beta-1}}{\Gamma(\beta)}\int_0^\infty e^{-ty}(1+y)^{\beta-1}k(\log(1+y))dy \tag{8.3.2}$$

*is the pdf of a GGC.*

**Proof.** We use that $P$ is the pgf of a GNBC iff $P(s+1) \in \hat{\mathcal{S}}$. For $\{p_j\}_0^\infty$ in (8.3.1), we have

$$P(z) = \int_0^\infty \sum_{j=0}^\infty (-1)^j \binom{-\beta}{j} (e^{-x}z)^j \, k(x)\, dx$$

$$= \int_0^\infty (1 - e^{-x}z)^{-\beta} k(x) dx = \int_0^\infty (1 - \tfrac{z}{y+1})^{-\beta} \, k(\log(1+y)) \frac{1}{1+y} dy \,.$$

Hence

$$P(s+1) = \int_0^\infty (y-s)^{-\beta}(1+y)^{\beta-1} k(\log(1+y)) dy = \int_0^\infty e^{st} g(t) dt$$

and the proof is completed. □

**Remark.** Substituting $y = 1/y'$ in (8.3.2), we see that $g(t)$ can be interpreted as the pdf of the product of a Gamma($\beta$,1)–variable and the rv $W$. It follows again by Theorem 6.2.1 that the earlier given sufficient condition is sufficient. □

There are two corollaries of Theorem 8.3.1. The first, of great importance, is immediate in view of the results in Section 5.2, whereas the second, less important, is a consequence of Theorem 6.2.5.

**Corollary 1.** *If* $(1+y)^{\beta-1} k(\log(1+y))$ $(= y^\beta f_{1/W}(y))$ *is proportional to the pdf of a GGC, maybe widened, then* $\{p_j\}_0^\infty$ *in (8.3.1) is a GNBC.*

**Corollary 2.** *If* $\beta \le 1$ *and* $(1+y)^{\beta-1} k(\log(1+y))$ *is logconcave, then* $\{p_j\}_0^\infty$ *in (8.3.1) is a GNBC.*

We now formulate and prove the main result of this section.

**Theorem 8.3.2.** *A pf of the form*

$$p_j = C \cdot (-1)^j \binom{-\beta}{j} \prod_{i=1}^n (t_i + j)^{-\gamma_i}, \quad j \in \mathbb{N}_0, \tag{8.3.3}$$

*is a GNBC if, for all* $i$, $\gamma_i \ge 1$, $t_i > 0$, *and* $\beta \le t_i + 1$.

**Proof.** Let $X = \sum_{i=1}^n X_i$, where the rv's $X_i \sim \text{Gamma}(\gamma_i, t_i)$ are independent. Clearly the appropriate function $k(x)$ is proportional to the pdf of $X$. Assume first that $\beta < t_i$ for all $i$. Then $e^{\beta x} k(x)$ is proportional to the pdf of $\tilde{X} = \Sigma \tilde{X}_i$, where

$\tilde{X}_i \sim \text{Gamma}(\gamma_i, t_i-\beta)$. This is obvious from the general relation $e^{\beta x}(k_1\star k_2\star \dots \star k_n) = e^{\beta x}k_1\star e^{\beta x}k_2\star \dots \star e^{\beta x}k_n$. It follows that $(1+y)^{\beta-1}k(\log(1+y))$ is proportional to the pdf of $\tilde{Y} = e^{\tilde{X}}-1$. By Theorem 6.2.3, $\tilde{Y} \sim \mathscr{T}$. Hence, by Corollary 1 of Theorem 8.3.1, $\{p_j\}_0^\infty$ is a GNBC.

One additional observation shows that it suffices to assume that $\beta \le t_i+1$. Consider the density $e^{\beta x}k_i(x)$, where $k_i(x) = x^{\gamma_i-1}e^{-t_i x}$. The density induced by this density and the transformation $x \to e^x-1 = y$ equals

$$(\log(1+y))^{\gamma_i-1}(1+y)^{-(1+t_i-\beta)},$$

and is HCM if $\beta \le 1+t_i$ and $\gamma_i \ge 1$; cf. Section 5.7. A simple generalization of Theorem 6.2.3 (based on a generalization of Theorem 6.2.2) then shows that $(1+y)^{\beta-1}k(\log(1+y))$ belongs to $\mathscr{T}_w$. Hence $\{p_j\}_0^\infty$ is a GNBC as desired. □

The restriction $\gamma_i \ge 1$ in Theorem 8.3.2 is annoying. There is a lot of numerical evidence supporting the following conjecture.

**Conjecture.** *A pf of the form (8.3.3) is a GNBC if, for all* $i$, $\gamma_i > 0$ *and* $0 < \beta \le t_i$ *(or, less restrictively,* $0 < \beta < t_i+\gamma_i$*).*

The conjecture in the weak version would in particular hold if $X \sim \mathscr{T} \Rightarrow e^X-1 \sim \mathscr{T}$. In one important special case the conjecture holds.

**Theorem 8.3.3.** *A pf of the form*

$$(8.3.4) \qquad p_j = C\cdot(-1)^j\binom{-\beta}{j}\cdot(t_1+j)^{-\gamma_1}\theta^j, \quad j \in \mathbb{N}_0,$$

*where* $\gamma_1 > 0$, $0 < \beta \le t_1$, *and* $0 < \theta \le 1$, *is a GNBC.*

**Proof.** The corresponding function $k(x)$ is proportional to the pdf of $X = X_1+a$, where $X_1 \sim \text{Gamma}(\gamma_1,t_1)$ and $e^{-a} = \theta$. It follows easily from Theorem 6.2.6 that $e^X-1 \sim \mathscr{T}$ and hence $\{p_j\}_0^\infty$ is a GNBC. □

For $\beta = t_1 = \gamma_1 = 1$, the distribution in (8.3.4) reduces to the Logarithmic series distribution. For $\beta = \theta = 1$ it may be called a *discrete Pareto distribution.* For general $\beta$ and $\theta = 1$ it is a discrete variant of the Beta distribution of the 2nd kind. Letting this latter discrete distribution sit on a fine lattice $\{j\cdot\epsilon, j \in \mathbb{N}_0\}$ and then letting $\epsilon \to 0$, we get the Beta distribution of the 2nd kind as a limit; cf. (ii) in Section 8.1.

Analogously all the distributions in $\mathscr{B}$ appear as limits of the pf's (8.3.3) with

$\gamma_i > 0$ and $0 < \beta \leq t_i$. These pf's would constitute the natural discrete $\mathscr{B}$–class, $\mathscr{B}_d$, if the conjecture above were true. However, there are also other candidates for being $\mathscr{B}_d$. The most obvious one is formed by the Po($\lambda$)–mixtures with $\lambda \sim \mathscr{B}$. Another possibility is to consider q–mixtures of the NB($\beta$,q)–distribution with $W = q/p \sim \mathscr{B}$. If $f_W(y) = C \cdot y^{\beta-1} \Pi_{i=1}^n (1+c_i y)^{-\gamma_i}$, then

$$f_q(q) = p^{-2} f_W(q/p) = C \cdot p^{\gamma-\beta-1} q^{\beta-1} \prod_{i=1}^{n} (p+c_i q)^{-\gamma_i} \quad (\gamma = \Sigma\, \gamma_i),$$

which is a kind of generalized Beta–density.

## 8.4 Complementary Results and Notes

I (***Poisson mixtures***). Consider some property a distribution on $(0, \infty)$ can have. Let $\mathscr{M}$ denote the corresponding class of distributions. It is a well–known fact that such properties are often inherited, in obvious discrete versions, by Po($\lambda$)–mixtures, $\lambda \sim \mathscr{M}$. As examples one can mention: *logconvexity, logconcavity, infinite divisibility, self–decomposability, and being* $PF_\infty$. In this chapter another property has been added to the list. We have also seen that a Po($\lambda$)–mixture is in $\mathscr{T}_d$ *iff* $\lambda \sim \mathscr{T}$. This equivalence is a rare exception. The "only if" does not hold for any of the five properties above. For example, for ID, Kallenberg (1976, p. 59) and others have provided counterexamples.

II ***(More about Poisson mixtures).*** Forst (1981) has shown that a pf $\{p_j\}_0^\infty$ is a Po($\lambda$)–mixture with $\lambda \sim$ ID **iff** $\{p_j\}_0^\infty$ is ID, $p_0 > 0$, and the r–sequence defined by (8.1.5) is such that $k! r_k = \int_{[0,\infty)} t^k \mu(dt)$ for some nonnegative measure $\mu(dt)$.

III (***Self–decomposability***). It can be verified by the help of VII in Section 2.5 that the pf $p_j = C \cdot \Pi_{j=1}^n (t_j+j)^{-\gamma_j}$ is SD provided that only $t_j \geq (\sqrt{5}-1)/2$ for all j. This result provides some support for the conjecture in Section 8.3.

**Notes.** The class of GNBC's was introduced in B1979c. The main goal there was to find a large class of explicit pf's in $\mathscr{T}_d$ but the goal was hardly achieved. Many of the results in this chapter are either new or improve considerably the earlier results.

The discrete Stable distributions in Example 8.2.1 were introduced by Steutel & van Harn (1979) in connection with discrete self–decomposability. The result in Example 8.2.2 is an improvement on Feller's (1966) result that passage times for simple random walks are ID. The nice Waring distribution example is due to Xekalaki (1983). Xekalaki & Panaretos (1988) obtained further results.

# 9. THE $\mathcal{T}_2$–CLASS OF DISTRIBUTIONS

Another extension of the class $\mathcal{T}$ is considered in this chapter. The new class is called the $\mathcal{T}_2$–class or the class of generalized convolutions of mixtures of Exponential distributions (GCMED). The latter name indicates how the distributions in $\mathcal{T}_2$ are generated.

In Section 9.1 the $\mathcal{T}_2$–distributions are studied via the mgf. Several examples are presented in Section 9.2. In Section 9.3 first passage time distributions for random walks in continuous time are shown to be in $\mathcal{T}_2$. In Section 9.4 densities of $\mathcal{T}_2$–distributions are the focus. Some further classes of distributions are briefly considered in Section 9.5.

## 9.1 Definition and Main Properties

The mgf $\varphi$ of a GGC satisfies $\varphi(0) = 1$ and the relation

$$\varphi'(s)/\varphi(s) = a + \int \frac{1}{t-s} U(dt)\ ; \quad a \geq 0, \quad U(dt) \geq 0.$$

(All integrals in this chapter are taken over $(0,\infty)$ unless otherwise indicated.) Since $\varphi(s) = \exp\{\log\varphi(s)\}$ and $(\log\varphi(s))'$ is absolutely monotone, it follows that $\varphi$ is the mgf of an ID distribution. Obviously the same conclusion holds if instead

(9.1.1) $$\varphi'(s)/\varphi(s) = a + \int \frac{1}{(t-s)^2} Q(dt)\ ; \quad a \geq 0, \ \ Q(dt) \geq 0\ .$$

A probability distribution on $\mathbb{R}_+ = [0,\infty)$ with mgf $\varphi$ satisfying (9.1.1) is called a $\mathcal{T}_2$–distribution; the index 2 refers to the exponent 2. Integrating (9.1.1) and using that $\varphi(0) = 1$, we get the equivalent more explicit definition:

**Definition.** *A $\mathcal{T}_2$–distribution is a probability distribution $F$ on $\mathbb{R}_+$ with mgf of the form*

*(9.1.2)* $$\varphi(s) = exp\{as + \int (\tfrac{1}{t-s} - \tfrac{1}{t}) Q(dt)\}, \quad s \leq 0,$$

*where $a \geq 0$ and the nonnegative measure $Q$ on $(0,\infty)$ satisfies $\int \frac{1}{t(1+t)} Q(dt) < \infty$.*

Here $a$ and Q are uniquely determined as will be verified later on. Obviously $\mathcal{T}_2$ is closed with respect to positive translation and change of scale, convolution and convolution roots. Not surprisingly, it is also closed with respect to weak limits. Let for

$F \in \mathcal{T}_2$ the measure $\nu$ on $[0,\infty]$ be defined by

$$\nu(dt) = \frac{1}{t(1+t)}Q(dt), \quad \nu(\{0\}) = 0, \quad \nu(\{\infty\}) = a. \tag{9.1.3}$$

**Theorem 9.1.1** ***(Closure theorem).*** *Let $F_n \in \mathcal{T}_2$, $n = 1, 2, \ldots$ If $F_n \to F$ weakly, then $F \in \mathcal{T}_2$ and $\nu_n \to \nu$ vaguely on $[0,\infty]$. Conversely, if $\nu_n$ converges vaguely on $[0,\infty]$ to some measure $\nu$ with $\nu(\{0\}) = 0$, then $F_n \to F$ weakly, where $F \in \mathcal{T}_2$ and is given by $\nu$. If $\nu(\{0\}) \neq 0$, then $F$ is a defective $\mathcal{T}_2$-distribution with total mass $e^{-\nu(\{0\})}$, i.e., $e^{\nu(\{0\})}F \in \mathcal{T}_2$.*

The proof is not hard and is left to the reader. Note that

$$\varphi_n(s) = \exp\{\int_{[0,\infty]} \frac{s(1+t)}{t-s}\,\nu_n(dt)\}, \quad s \le 0,$$

and that the function $[0,\infty] \ni t \to \frac{s(1+t)}{t-s}$, $s < 0$, is negative and continuous with value $-1$ at $t = 0$ and value $s$ at $t = \infty$.

There is a simple characterization of the $\mathcal{T}_2$–distributions in terms of the Lévy measure.

**Theorem 9.1.2.** *A distribution on $\mathbb{R}_+$ belongs to $\mathcal{T}_2$ iff it is ID and the Lévy measure has a completely monotone (CM) density $\ell$. In fact, $\ell(y) = \int e^{-ty}Q(dt)$.*

**Proof.** The theorem is obvious from the relation

$$\int (t-s)^{-2}Q(dt) = \int \left(\int y\cdot\exp\{-y(t-s)\}dy\right) Q(dt) = \int e^{sy}\left(y \int e^{-ty}Q(dt)\right) dy. \quad \square$$

Since $\ell(y)$ is CM when $y\ell(y)$ is, it follows that $\mathcal{T} \subset \mathcal{T}_2$; cf. Theorem 3.1.1. We also see that a $\mathcal{T}_2$–distribution is not necessarily self–decomposable. The GGC's are characterized by the Pick function property of $\varphi'(s)/\varphi(s)$. There are two corresponding results for the $\mathcal{T}_2$–distributions.

**Theorem 9.1.3.** *A distribution on $\mathbb{R}_+$ belongs to $\mathcal{T}_2$ iff its mgf $\varphi$ is analytic and zero–free in $\mathbb{C}\backslash[0,\infty)$ and*

*a)* $\Im m[\log(\varphi(s))] = \arg[\varphi(s)] \ge 0$ *for* $\Im m\, s > 0$

*or*

*b)* $\Im m[\frac{1}{s}\log(\varphi(s)] \ge 0$ *for* $\Im m\, s > 0$.

This theorem is a rather immediate consequence of the definition of $\mathcal{T}_2$ and Lemmata 2.4.1, 2.4.2, and 2.4.3 in Section 2.4 on Pick functions. Note that for a $\mathcal{T}_2$–distribution

$$\frac{1}{s}\log(\varphi(s)) \;=\; a + \int \frac{1}{t-s}\frac{1}{t}\,Q(dt).$$

Theorem 9.1.3 shows that for a $\mathcal{T}_2$–distribution the measure Q is uniquely determined. In fact, by Theorem 2.4.1,

$$\int_{(c,d)} Q(dt) \;=\; \lim_{\eta\downarrow 0} \frac{1}{\pi} \int_{(c,d)} \Im m[\log(\varphi(x+i\eta))]dx\,.$$

if $Q(\{c\}) = Q(\{d\}) = 0$. If, for instance, $\varphi$ is continuous up to and on the cut, then Q has a density q(t) given by $q(t) = \pi^{-1}\arg[\varphi(t+i\cdot 0)]$, $t > 0$.

As seen earlier, $\mathcal{T} \subset \mathcal{T}_2$. This result follows also from Theorem 9.1.3 since, for a GGC,

$$\Im m[\log(\varphi(s)] = a\cdot\Im m\, s \;+ \int \Im m[\log(\tfrac{t}{t-s})]\;U(dt)$$

and

$$\Im m[\log(\tfrac{t}{t-s})] = \arg[\tfrac{t}{t-s}] \;\in\; (0,\pi) \quad\text{in}\quad \Im m\, s > 0.$$

The class of mixtures of Exponential distributions (MED) is also a subclass of $\mathcal{T}_2$. In fact, for the mgf $\varphi$ of a (non–degenerate) MED, $\Im m[\varphi(s)] > 0$ for $\Im m\, s > 0$ (cf. Theorem 2.4.2) and hence

$$\Im m[\log(\varphi(s)] = \arg[\varphi(s)] \in (0,\pi)\;, \quad \Im m\, s > 0.$$

The following two theorems give more information about the relations between $\mathcal{T}$ and $\mathcal{T}_2$ and the class of MED's and $\mathcal{T}_2$.

**Theorem 9.1.4.** *The $\mathcal{T}$–class is formed by those distributions in $\mathcal{T}_2$ for which the measure $Q$ has an increasing density $q(t)$. In fact, if $U(dt)$ denotes the Thorin measure, then $q(t) = U(t) = \int_{(0,t]} U(d\tilde{t})$.*

**Theorem 9.1.5.** *The class of MED's is formed by those distributions in $\mathcal{T}_2$ for which the left–extremity $a$ is zero and $Q$ has a density $q(t)$ such that $q(t) \le 1$.*

**Proof of Theorem 9.1.4.** Let $\varphi \in \hat{\mathcal{T}}$. Then $\varphi(s) = e^{\zeta(s)}$, where

(9.1.4) $$\zeta(s) = as + \int \log(\tfrac{t}{t-s})U(dt)\,.$$

Here $\int_{(0,1]} |\log t| U(dt) < \infty$ and $\int_{(1,\infty)} t^{-1}U(dt) < \infty$ and hence, as is easily shown, $t^{-1}U(t) \to 0$ as $t \to \infty$ and $|\log t| U(t) \to 0$ as $t \to 0$. An integration by parts yields

$$\zeta(s) = as + \int \left(\frac{1}{t-s} - \frac{1}{t}\right) q(t)dt \,, \tag{9.1.5}$$

where $q(t) = U(t)$. Thus $\varphi \in \hat{\mathcal{T}}_2$ and the Q–measure has an increasing density. Conversely, if this holds, then, by also the condition $\int \frac{q(t)}{t(1+t)}dt < \infty$, it easily follows that $t^{-1}q(t) \to 0$ as $t \to \infty$ and $|\log t| q(t) \to 0$ as $t \to 0$. An integration by parts of (9.1.5) then gives (9.1.4) with $U(dt) = q(t)dt$. This ends the proof. □

**Proof of Theorem 9.1.5.** Let $\varphi$ be the mgf of a non–degenerate MED. Then, as was found earlier, $\varphi \in \hat{\mathcal{T}}_2$ and, for $\Im m\, s > 0$, $\Im m[\log \varphi(s)] \in (0,\pi)$. Hence

$$\int_{(c,d)} Q(dt) = \lim_{\eta \downarrow 0} \frac{1}{\pi} \int_{(c,d)} \Im m[\log \varphi(x+i\eta)]dx \le (d-c)$$

if $Q(\{c\}) = Q(\{d\}) = 0$. It follows that $Q$ is absolutely continuous with a density bounded by 1. Conversely, let $\varphi \in \hat{\mathcal{T}}_2$ with $a = 0$ and $Q(dt) = q(t)dt$, where $0 \le q(t) \le 1$ and $q(t) \not\equiv 0$. As $\log \varphi(s) = \int \left(\frac{1}{t-s} - \frac{1}{t}\right) q(t)dt$, it follows that, for $\Im m\, s > 0$,

$$0 < \Im m[\log \varphi(s)] = \int \frac{\Im m s}{|t-s|^2} q(t)dt \le \int \frac{\Im m s}{|t-s|^2}dt = \frac{\pi}{2} + \arctan\left(\frac{\Re e\, s}{\Im m s}\right) < \pi \,.$$

Thus $\arg[\varphi(s)] \in (0,\pi)$ for $\Im m\, s > 0$. Consequently, $\Im m[\varphi(s)] > 0$ for $\Im m\, s > 0$ and Theorem 2.4.2 shows that $\varphi$ is the mgf of an MED. □

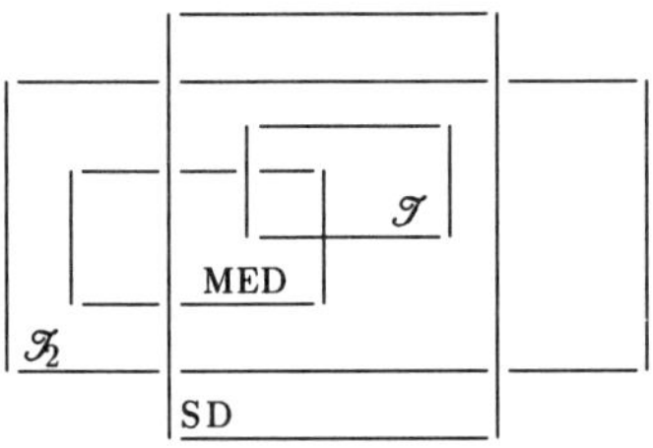


Figure 9.1.1. *Venn diagram for some classes of distributions*

For a finite MED the mgf has the form

$$\varphi(s) = \sum_{k=1}^{n} m_k \frac{t_k}{t_k - s} \,,$$

where $m_k, t_k > 0$ and the sum $\Sigma\, m_k$ is 1 or possibly $\leq 1$ (defective case). The rational function $\varphi(s)$, $s > 0$, is real and strictly increasing from $-\infty$ to $\infty$ between two consecutive poles. The zeros and the poles are thus interlacing. We get

$$q(t) = \frac{1}{\pi}\cdot \arg[\varphi(t+i\cdot 0)] = \begin{cases} 1 & \text{if } \varphi(t) < 0 \\ 0 & \text{if } \varphi(t) > 0\,. \end{cases}$$

Since there are only a finite number of zeros and poles of $\varphi$, the density $q(t)$ is fully defined by the above. In particular, if

$$\varphi(s) = m_1 + m_2\cdot\frac{t_2}{t_2-s} \qquad (t_1 = \infty),$$

corresponding to a mixture of a one–point distribution at 0 and an Exponential distribution, then

$$\text{(9.1.6)} \qquad q(t) = \begin{cases} 1 & \text{if } t_2 < t < \frac{m_1+m_2}{m_1}\, t_2 \\ 0 & \text{elsewhere} \end{cases}$$

It is now rather obvious that any $\mathcal{T}_2$–distribution is a weak limit of a sequence of finite convolutions of such *elementary mixtures*; the measure $\nu$ defined by (9.1.3) is just approximated by measures of the form $\frac{1}{t(t+1)}\Sigma_{i=1}^{N} q_i(t)dt$, where each $q_i(t)$ is of the form (9.1.6). Of course, a finite MED is a finite convolution of elementary mixtures. Conversely, a weak limit of a sequence of convolutions of MED's is obviously in $\mathcal{T}_2$. The $\mathcal{T}_2$–distributions are therefore also called generalized convolutions of MED's. Originally the class $\mathcal{T}_2$ was introduced to get a class that is closed with respect to weak limits and convolutions and that contains the MED's as well as the GGC's.

Contrary to the case for GGC's, there is a very simple composition theorem for $\mathcal{T}_2$–distributions.

**Theorem 9.1.6** ***(Composition theorem).*** *Let* $e^{\zeta(s)} \in \hat{\mathcal{T}}_2$ *. Then* $\varphi(\zeta(s)) \in \hat{\mathcal{T}}_2$ *whenever* $\varphi \in \hat{\mathcal{T}}_2$*. Moreover,* $\varphi(\zeta(s))$ *is the mgf of an MED whenever* $\varphi$ *is.*

**Proof.** Let $\varphi \in \hat{\mathcal{T}}_2$. We know that $\varphi(\zeta(s))$ is the mgf of $X(T)$, where $X(\tau)$, $\tau \geq 0$, is a Lévy process with mgf $e^{\tau\zeta(s)}$ and $T$ is a random time with mgf $\varphi$. Since $\zeta(s)$ is of the form $\zeta(s) = as + \int (\frac{1}{t-s} - \frac{1}{t})Q(dt)$, it maps (unless $\zeta(s) \equiv 0$) the open upper (lower) half–plane into itself and the negative real line into itself. The function $\varphi(\zeta(s))$ is therefore well–defined and analytic in $\mathbb{C}\backslash[0,\infty)$ and zero–free there. Since, by Theorem 9.1.3, $\Im m[\log\varphi(s)] \geq 0$ for $\Im m\, s > 0$, it follows that also $\Im m[\log\varphi(\zeta(s))] \geq 0$ for

$\Im m\, s > 0$ and hence $\varphi(\zeta(s)) \in \hat{\mathcal{T}}_2$ by again Theorem 9.1.3. If $\varphi$ is instead the mgf of an MED, then $\Im m[\varphi(s)]$ is nonnegative for $\Im m\, s > 0$ and hence so is $\Im m[\varphi(\zeta(s))]$; this gives the last part of the theorem. □

**Corollary.** *If* $e^{\zeta(s)} \in \hat{\mathcal{T}}$, *then* $\varphi(\zeta(s)) \in \hat{\mathcal{T}}_2$ *whenever* $\varphi \in \hat{\mathcal{T}}$.

We end this section by giving a real characterization of the LT $\phi(s)$ $(= \varphi(-s))$ of a $\mathcal{T}_2$–distribution. It is not as useful as the corresponding characterization for GGC's.

**Theorem 9.1.7** ***(Real characterization).*** *A probability distribution on* $\mathbb{R}_+$ *with LT* $\phi$ *belongs to* $\mathcal{T}_2$ *iff, for each* $u > 0$, $\dfrac{\log(\phi(u/v)/\phi(uv))}{v - v^{-1}}$ *is a CM function of* $w = v + v^{-1}$.

**Proof.** For a $\mathcal{T}_2$–distribution we immediately get

$$\frac{\log(\phi(u/v)/\phi(uv))}{v - v^{-1}} = au + \int \frac{u}{t^2+u^2 + u(v+v^{-1})} Q(dt)$$

which is CM in w. Turning to the converse part, we set $\psi(s) = \frac{1}{-s}\cdot\log\phi(s)$. Now

$$\frac{v\psi(uv) - v^{-1}\psi(u/v)}{v - v^{-1}}$$

is CM in w. In view of Theorem 6.1.3 and since $s\psi(s) \to 0$ as $s \downarrow 0$, it follows that

$$\psi(s) = a + \int_{[0,\infty)} \frac{1}{t+s} U(dt), \quad a \geq 0, \; U(dt) \geq 0,$$

and hence

$$\log\phi(s) = -as - U(\{0\}) + \int \left(\frac{1}{t+s} - \frac{1}{t}\right)Q(dt),$$

where $Q(dt) = tU(dt)$. Since $\phi(0) = 1$, necessarily $U(\{0\}) = 0$, and hence $\phi$ is the LT of a $\mathcal{T}_2$–distribution. □

**Remark.** By Theorem 6.1.3 we have that the LT $\phi$ of an MED is characterized by complete monotonicity of $\dfrac{v\phi(uv) - v^{-1}\phi(u/v)}{v - v^{-1}}$ as a function of w. □

## 9.2 Examples of $\mathcal{T}_2$–distributions

Some examples of genuine $\mathcal{T}_2$–distributions are given in this section. Further important examples appear in Section 9.3 and in Chapter 10 in the context of shot–noise processes.

**Example 9.2.1** ***(Compound Poisson distribution).*** Consider $X = \Sigma_{i=1}^{N} Y_i$, where here $N \sim Po(\lambda)$ and $Y_i \sim Exp(t_0)$, i = 1, 2, ..., and these rv's are independent. Then

$$\varphi_X(s) = \exp\{\lambda\cdot(\varphi_{Y_1}(s)-1)\} = \exp\{\lambda t_0(\tfrac{1}{t_0-s} - \tfrac{1}{t_0})\}.$$

Thus $X \sim \mathcal{G}_2$ ; the Q–measure is degenerate at $t_0$. Obviously $X = 0$ with probability $e^{-\lambda}$. For $t_0 = 1$ we have

$$f_X(x) = \sum_{k=1}^{\infty} \frac{x^{k-1}e^{-x}}{(k-1)!}\cdot\frac{\lambda^k e^{-\lambda}}{k!}, \quad x > 0.$$

This density is increasing in a neighbourhood of zero if $\lambda > 2$; in this case the distribution is not unimodal because of the mass at 0.

More generally, $X \sim \mathcal{G}_2$ if the common distribution of the variables $Y_i$ is an MED with mgf $\varphi_{Y_i}(s) = \int_{(0,\infty]} \frac{t}{t-s} M(dt)$, where M has total mass 1. Obviously $Q(dt) = \lambda t M(dt)$ on $(0,\infty)$. A given $\mathcal{G}_2$–distribution with left–extremity 0 can be represented in this way iff $\int t^{-1}Q(dt) < \infty$; $\lambda$ is chosen equal to (or larger than) $\int t^{-1}Q(dt) = \int L(dy)$, where L is the Lévy measure. If $\int L(dy) = \infty$, then by e.g. Theorem 5.5.7 in Lukacs (1970), the distribution is absolutely continuous. □

**Example 9.2.2** ***(Non–central $\chi^2$–distribution).*** Let $Z_i \sim N(\mu_i,1)$, i = 1, ..., n, be independent rv's. Then $X = \Sigma Z_i^2$ has a non–central $\chi^2$–distribution. It is not hard to show that its mgf is given by

$$\varphi_X(s) = (1-2s)^{-n/2}\cdot\exp\{\tfrac{\lambda s}{1-2s}\} = (1-2s)^{-n/2}\cdot\exp\{\tfrac{\lambda}{2}(\tfrac{1}{1-2s} - 1)\},$$

where $\lambda = \Sigma \mu_i^2$. The last main factor is the mgf of $\Sigma_{j=1}^{N} Y_j$, where $N \sim Poisson(\lambda/2)$ and $Y_j \sim Exp(1/2)$. Hence the non–central $\chi^2$–distribution belongs to $\mathcal{G}_2$. □

**Example 9.2.3** ***(Logarithm of Beta variable).*** Let $X = -\log Y$, where $Y \sim Beta(\alpha,\beta)$. We may think of Y as a random probability. For instance, it may be the probability that a seed sown in a spot generates a seedling; Y varies from spot to spot. The logarithmization transforms the multiplicative structure of probabilities into an additive structure. Now

$$\varphi_X(s) = \mathbf{E}[Y^{-s}] = \int_0^1 y^{-s}\frac{1}{B(\alpha,\beta)}y^{\alpha-1}(1-y)^{\beta-1}dy = \frac{\Gamma(\alpha+\beta)\Gamma(\alpha-s)}{\Gamma(\alpha)\Gamma(\alpha+\beta-s)}.$$

Let $\psi(z) = \Gamma'(z)/\Gamma(z) = -C + \Sigma_{n=1}^{\infty}(\frac{1}{n} - \frac{1}{n-1+z})$, where C is Euler's constant. We get

$$\text{(9.2.1)}\qquad \frac{\varphi_X'(s)}{\varphi_X(s)} = \sum_{n=1}^{\infty}\Big(\frac{1}{n-1+\alpha-s} - \frac{1}{n-1+\alpha+\beta-s}\Big) = \sum_{n=1}^{\infty}\int_{n-1+\alpha}^{n-1+\alpha+\beta}\frac{1}{(t-s)^2}\,dt\,.$$

Thus $X \sim \mathcal{G}_2$ with $a = 0$ and $Q(dt) = q(t)dt$, where

$$q(t) = \sum_{n=1}^{\infty} 1_{(n-1+\alpha,\ n-1+\alpha+\beta)}(t)\,.$$

Hence $X \sim$ MED iff $\beta \le 1$. Moreover, as is easily seen from (9.2.1), $X \sim \mathcal{G}$ iff $\beta \in \mathbb{N}$. We get also an expression for the Lévy density $\ell$ from (9.2.1). We have

$$y\ell(y) = \sum_{n=1}^{\infty}\Big(\exp\{-(n-1+\alpha)y\} - \exp\{-(n-1+\alpha+\beta)y\}\Big) = e^{-\alpha y}(1-e^{-\beta y})/(1-e^{-y}).$$

It can be verified that this function is decreasing, i.e. X is self–decomposable (SD), iff $2\alpha + \beta \ge 1$. □

**Example 9.2.4** ***(Autoregressive process).*** Consider a sequence $\{X_n\}_0^{\infty}$ of identically distributed SD random variables and defined by

$$X_n = cX_{n-1} + \epsilon_n, \quad n = 1, 2, \dots,$$

where $0 < c < 1$ and the random innovation $\epsilon_n$ is independent of $X_0, X_1, \dots, X_{n-1}$. Now assume that $X_n \sim \mathcal{G}$. We shall see that then $\epsilon_n \sim \mathcal{G}_2$. Omitting the index n, we have $\varphi_\epsilon(s) = \varphi_X(s)/\varphi_X(cs)$ and hence, with obvious notation,

$$\text{(9.2.2)}\qquad \begin{aligned}\frac{\varphi_\epsilon'(s)}{\varphi_\epsilon(s)} &= \frac{\varphi_X'(s)}{\varphi_X(s)} - c\,\frac{\varphi_X'(cs)}{\varphi_X(cs)} = a(1-c^2)s + \int\Big(\frac{1}{t-s} - \frac{c}{t-cs}\Big)U(dt)\\ &= a(1-c^2)s + \int\frac{1}{(t-s)^2}\Big(U(t) - U(ct)\Big)dt\,.\end{aligned}$$

The last equality is obtained by essentially an integration by parts. Since $q(t) = U(t) - U(ct)$ is nonnegative, $\epsilon \sim \mathcal{G}_2$.

There is also a converse: If X is SD and if, moreover, for each c, $0 < c < 1$, we have $X \overset{d}{=} cX + \epsilon$ with $\epsilon \sim \mathcal{G}_2$, then $X \sim \mathcal{G}$. To see this, let $\ell$ be the Lévy density corresponding to X. Then

$$\varphi_\epsilon'(s)/\varphi_\epsilon(s) = \int e^{sy}\Big(y\ell(y) - \tfrac{y}{c}\ell(\tfrac{y}{c})\Big)dy,$$

i.e., the Lévy density corresponding to $\epsilon$ is $\ell_\epsilon(y) = \ell(y) - \frac{1}{c}\ell(\frac{y}{c})$. We have to show that if $\ell_\epsilon$ is CM for each c, then so is $r(y) = y\ell(y)$; cf. Theorems 9.1.2 and 3.1.1. However, since CM is preserved under pointwise limits, this follows from the fact that, as $c \uparrow 1$,

$$\frac{1}{(c^{-1}-1)}\cdot \ell_\epsilon(y) = \frac{r(y) - r(y/c)}{(c^{-1}-1)\cdot y} \to -r'(y).$$

It may be added that if $X \sim \mathcal{T}$ and the corresponding U–measure has a density $U'(t)$, then $\epsilon \sim \mathcal{T}$ for all c iff $t\cdot U'(t)$ is increasing. □

**Example 9.2.5** ***(Size–biased sampling).*** Let $f \in \mathcal{T}$ and consider the new pdf $g(x) = C\cdot xf(x)$ corresponding to size–biased sampling from f; it is assumed that $\int xf(x)dx < \infty$. Then $g \in \mathcal{T}_2$. In fact, g is the convolution of $f \in \mathcal{T}$ and an MED; cf. Example 3.2.6. Hence if $X \sim f$ and $\tilde{X} \sim g$, then $\tilde{X} \overset{d}{=} X + \epsilon$, where $\epsilon$ is independent of X and $\epsilon \sim \text{MED} \subset \mathcal{T}_2$. □

**Example 9.2.6** ***(Curious decomposition).*** Now let $\tilde{X}$ instead stem from $C\cdot e^{-\theta x}f(x)$, $\theta > 0$. Then, for $X \sim f \in \mathcal{T}$, we have, as is easily verified, the curious decomposition $X \overset{d}{=} \tilde{X} + \epsilon$, where $\epsilon$ is independent of $\tilde{X}$ and $\epsilon$ has a $\mathcal{T}_2$–distribution with Q–measure with density $q(t) = U(t) - U(t-\theta)$, $t > 0$; cf. (9.2.2) in Example 9.2.4. There is a converse result as well; see Berg & Forst (1982). □

## 9.3 First Passage Time Distributions as $\mathcal{T}_2$–distributions

In this section we shall see that $\mathcal{T}_2$ contains every first passage time distribution for a random walk on $\mathbb{Z}$ in continuous time. Moreover, every GGC is at least a limit of such passage time distributions.

We consider a random walk on $\mathbb{Z}$ in continuous time with transition intensities $\lambda_k$ $(k \to k+1)$ and $\mu_k$ $(k \to k-1)$. A jump from state k occurs after a random time $\sim \text{Exp}(\lambda_k+\mu_k)$. The transition $k \to k+1$ has probability $\lambda_k/(\lambda_k+\mu_k)$. Let $T_{mn}$ denote the first passage time from state m to state n. Then, if $m < n$,

$$T_{mn} = T_m^+ + T_{m+1}^+ + \dots + T_{n-1}^+,$$

where $T_k^+ = T_{k,k+1}$. The random times in the sum are independent by the strong Markov property for a random walk. If n is not reached, then $T_{mn} = \infty$. We let $T_{mn} = \infty$ also if the walk reaches $-\infty$ in finite time. The following simple result holds.

**Theorem 9.3.1.** *The distribution of* $T_k^+$ *is an MED, maybe defective. Hence* $T_{mn}$ *has a* $\mathcal{J}_2$*–distribution, maybe defective.*

**Proof.** Let $\varphi_k^+$ denote the mgf of $T_k^+$. We easily get

$$\varphi_k^+(s) = \frac{\lambda_k+\mu_k}{\lambda_k+\mu_k-s}\cdot\Big(\frac{\lambda_k}{\lambda_k+\mu_k}+\frac{\mu_k}{\lambda_k+\mu_k}\cdot\varphi_{k-1}^+(s)\cdot\varphi_k^+(s)\Big),$$

and hence

$$\varphi_k^+(s) = \frac{\lambda_k}{\lambda_k+\mu_k-s-\mu_k\varphi_{k-1}^+(s)}.$$

If $\varphi_{k-1}^+(s)$ is the mgf of an MED, maybe defective, then $\Im m[\varphi_{k-1}^+(s)] \geq 0$ for $\Im m\, s > 0$. Since the conjugate of the denominator above then has nonnegative imaginary part in the upper half–plane, it follows that $\Im m[\varphi_k^+(s)] \geq 0$ for $\Im m\, s > 0$. Hence, by a simple variant of Theorem 2.4.2, the distribution of $T_k^+$ is an MED, maybe defective. By *induction* it follows that this conclusion is true whenever $\lambda_j = 0$ for some $j < k$. Since the distribution of $T_k^+$ is only slightly affected if the original walk is changed so that $\lambda_j = 0$ for some very large negative $j$, it follows that, in full generality, $T_k^+ \sim$ MED. Hence $T_{mn} \sim \mathcal{J}_2$. □

There is another decomposition of $T_{mn}$ which gives more information on the distribution of $T_{mn}$. We have, as noted by Keilson (1981),

$$T_{mn} = T^{\#n}_{m,m+1} + {}^{m}T_{m+1,n}\,,$$

where $T^{\#n}_{m,m+1}$ is the upwards last exit time of state $m$ before state $n$ is reached and ${}^{m}T_{m+1,n}$ is the subsequent time to reach $n$ conditional on that state $m$ is not visited again. These two random times are independent as can be seen upon some reflection. Assuming that $\lambda_m, \lambda_{m+1}, \ldots, \lambda_{n-1}$ are positive, we have:

**Theorem 9.3.2.** *The distribution of* $T^{\#n}_{m,m+1}$ *is an MED, maybe defective, and the distribution of* ${}^{m}T_{m+1,n}$ *is a convolution of* $n-m-1$ *Exponential distributions.*

Only the first part of the theorem is proved here; see Keilson (1981) for a more complete proof.

**Proof.** Let $m = 0$. We have

$$T^{\#n}_{01} = T_{01} + \sum_{j=1}^{N}\Big({}^{n}T_{10}^{(j)} + T_{01}^{(j)}\Big),$$

where $N$ is a geometrically distributed rv representing the number of returns (possibly zero) to state 0 from state 1 before state n is reached and where ${}^{n}T_{10}^{(j)}$ denotes a passage time from state 1 to state 0 conditional on that state n is not visited. What $T_{01}^{(j)}$ stands for should be obvious. Some reflection shows that ${}^{n}T_{10}^{(j)} \sim \mathrm{MED}$ by essentially Theorem 9.3.1. Now

$$\varphi_{01}^{\#n}(s) = \varphi_{01}(s)\cdot\frac{1-q}{1-q\cdot{}^{n}\varphi_{10}(s)\cdot\varphi_{01}(s)};$$

here q is a parameter for the distribution of N. We know that $\Im m[\varphi_{01}(s)] \geq 0$ and $\Im m[{}^{n}\varphi_{10}(s)] \geq 0$ for $\Im m\, s > 0$. It follows easily that the imaginary part of $\varphi_{01}(s)$ times the conjugate of the denominator above is nonnegative for $\Im m\, s > 0$. Hence, $\Im m[\varphi_{01}^{\#n}(s)] \geq 0$ for $\Im m\, s > 0$ and thus, by Theorem 2.4.2 again, $T_{01}^{\#n} \sim \mathrm{MED}$. □

By Theorem 9.3.2, the distribution of $T_{0n}$ is a convolution of an MED and a convolution of n–1 Exponential distributions. This result has a converse:

*Any distribution which is the convolution of an MED and n–1 Exponential distributions is at least a limit distribution of $T_{0n}$ for a sequence of appropriately chosen random walks.*

In fact, as mentioned by Steutel (1970), cf. B1981 and its addendum, any mixture of a finite number of non–degenerate Exponential distributions corresponds to $T_{01}$ for some random walk. The result then follows by setting $\mu_1 = \mu_2 = \ldots = \mu_n = 0$ and choosing $\lambda_1, \lambda_2, \ldots, \lambda_{n-1}$ appropriately.

The possible distributions of $T_{0n}$ can therefore *essentially* be described as those distributions for which the mgf has the form

$$\varphi(s) = \exp\{c + \int (\tfrac{1}{t-s} - \tfrac{1}{t})(q(t) + V(t))dt\},$$

where $c \leq 0$, $0 \leq q(t) \leq 1$, and V(t) is an increasing piecewise constant function with n–1 jumps of height 1. If c = 0, the distribution is non–defective. Although q(t) + V(t) is unique, the components q and V are not.

Now let U(dt) be the Thorin measure of a GGC and let $U(\infty) \leq n$. By a graph it is easily seen that U(t) can be represented as U(t) = q(t) + V(t), where $0 \leq q(t) \leq 1$ and V(t) is as above with at most n–1 jumps. By some reflection, we conclude:

*Any GGC is at least a limit of first passage time distributions for random walks.*

By a limit reasoning it follows that also for diffusion processes on $\mathbb{R}$ first passage time distributions belong to $\mathcal{T}_2$. Much more about this can be found in papers by Kent

(1980, 1982). Random walks in discrete time have been treated by Kent & Longford (1983), Sumita & Masuda (1985), Masuda (1988) and in B1991. Pitman & Yor (1981) give many results about infinite divisibility for the Bessel process.

### 9.4 On Densities of Distributions in $\mathscr{T}_2$

Example 9.2.1 showed essentially that a $\mathscr{T}_2$–distribution with left–extremity $a = 0$ is absolutely continuous for $x > 0$ but has the mass $\exp\{-\int t^{-1}Q(dt)\}$ (which may be 0) at $x = 0$. There is also a Gamma mixture representation for some $\mathscr{T}_2$–distributions.

**Theorem 9.4.1.** *A $\mathscr{T}_2$–distribution with $a = 0$ and spectral density $q(t)$ such that esssup $q(t)$ is finite can be represented as a mixture of Gamma$(N,\cdot)$–distributions, where $N$ is the smallest integer $\geq$ esssup $q(t)$.*

**Proof.** As $q(t)/N \leq 1$, the distribution is a convolution of N identical MED's. A mixture of Gamma$(\beta_1,\cdot)$–distributions convolved with a mixture of Gamma$(\beta_2,\cdot)$–distributions is a mixture of Gamma$(\beta_1+\beta_2,\cdot)$–distributions. By induction, the desired result follows. □

**Remark.** It is not true that in general a $\mathscr{T}_2$–distribution with $a = 0$ and $q(t) \leq \beta$ is a mixture of Gamma$(\beta,\cdot)$ distributions. A counterexample is provided by a $\mathscr{T}_2$–distribution with mgf $\varphi(s) = (p + (1-p)\frac{1}{1-s})^{\beta}$, $0 < p < 1$, $\beta < 1$, for which $q(t) = \frac{1}{\pi}\cdot\arg[\varphi(t+i\cdot 0)] = \beta$ if $1 < t < 1/p$ and else 0. It is not hard to see that for $\psi(s) = \int_{(0,\infty]}(\frac{t}{t-s})^{\beta}M(dt)$, the function $\frac{1}{\pi}\cdot\arg[\psi(t+i\cdot 0)]$ cannot equal $q(t)$. □

We now turn to an analogue of the class $\mathscr{B}$ so successful for GGC's. It is natural to define $\mathscr{B}_2$ as the class of distributions with pdf's of the form

$$f(x) = C\cdot x^{\beta-1}h(x),\ \ x > 0, \tag{9.4.1}$$

where $\beta > 0$ and $h \in \check{\mathscr{T}}_2$ (i.e., h is the LT of a $\mathscr{T}_2$–distribution), or pointwise limits of functions of this form. Obviously $\mathscr{B} \subset \mathscr{B}_2$. Alternatively, $\mathscr{B}_2$ can be defined by the generating class of functions

$$f(x) = C\cdot x^{\beta-1}\prod_{i=1}^{n}\left(p_i + (1-p_i)\frac{1}{1+c_i x}\right), \tag{9.4.2}$$

where $\beta > 0$, $0 \leq p_i < 1$, and $c_i > 0$. The factors in the product are LT's of elementary

MED's. The class $\mathscr{B}_2$ shares two important properties with $\mathscr{B}$.

**Theorem 9.4.2.** *If* $X \sim \mathscr{B}_2$, *then* $1/X \sim \mathscr{B}_2$.

**Proof.** It suffices to consider the case when $f_X$ has the form (9.4.2). Then

$$f_{X^{-1}}(x) = x^{-2}f_X(1/x) = C\cdot x^n\cdot x^{-\beta-1}\prod_{i=1}^n\left(p_i\frac{1}{x} + (1-p_i)\frac{1}{x+c_i}\right)$$

$$= \lim_{\delta\to 0} C\cdot x^n\cdot(x+\delta)^{-\beta-1}\prod_{i=1}^n\left(p_i\frac{1}{x+\delta} + (1-p_i)\frac{1}{x+c_i}\right).$$

Each factor to the right of $x^n$ is proportional to the LT of a $\mathscr{T}_2$–distribution and the result follows. □

A second property is given as a rather general theorem.

**Theorem 9.4.3.** *Let* $\mathscr{P}$ *be a class of probability distributions on* $[0,\infty)$ *which is scale–invariant and closed with respect to weak limits. Let* $\mathscr{B}(\mathscr{P})$ *be the class of pdf's on* $(0,\infty)$ *which are of the form* $f(x) = C\cdot x^{\beta-1}h(x)$, *where* $\beta > 0$ *and* $h \in \breve{\mathscr{P}}$, *or are pointwise limits of functions of that form. If* $X$ *is a positive rv such that* $C\cdot x^n e^{-\epsilon x}f_X(x) \in \mathscr{P}$ *for all* $n \in \mathbb{N}_0$ *and all* $\epsilon > 0$, *then* $1/X \sim \mathscr{B}(\mathscr{P})$. *In particular, if* $\mathscr{P} = \mathscr{T}_2$, *then* $1/X \sim \mathscr{B}_2 = \mathscr{B}(\mathscr{T}_2)$ *and hence* $X \sim \mathscr{B}_2$.

**Sketch of Proof.** The proof is based on the arguments used in the proof of Theorem 5.3.1 and is only sketched. Let $Y_n \sim \text{Gamma}(n,1)$ and consider $Y_n/(nX)$. Its pdf is given by (5.3.1). It suffices to show that $Y_n/(nX) \sim \mathscr{B}(\mathscr{P})$ since $Y_n/n \to 1$ as $n \to \infty$. Now $f_n = \lim_{\epsilon\to 0} f_{n,\epsilon}$, with $f_{n,\epsilon}$ as in the proof of Theorem 5.3.1. Since $f_{n,\epsilon} \in \mathscr{B}(\mathscr{P})$, the first part of the theorem follows. The second part is a consequence of Theorem 9.4.2. □

A simple example of a pdf in $\mathscr{B}_2$ (but not in $\mathscr{B}$) is provided by $f(x) = C\cdot(1+cx)x^{\beta-1}e^{-x}$. Its mgf equals $\varphi(s) = C\cdot\Gamma(\beta)\cdot(1-s)^{-\beta}(1+\frac{c\beta}{1-s})$ and hence $f \in \mathscr{T}_2$. Another similar example is now given.

**Example 9.4.1** ***(Birnbaum–Saunders' distribution).*** Consider the inverse Gaussian distribution with pdf and mgf

$$f(x) = C\cdot x^{-3/2}\exp\{-c_1x - c_2x^{-1}\}, \qquad \varphi(s) = \exp\{c_3(1-\sqrt{1-c_4s})\},$$

where $c_3 = 2\sqrt{c_1c_2}$ and $c_4 = 1/c_1$. We now form a new pdf g by

$$g(x) = C'\cdot(p + q\sqrt{c_1/c_2}\,x)\cdot f(x), \qquad q = 1-p,\ 0 \le p \le 1.$$

This distribution, an *inverse Gaussian mixture distribution*, was introduced and studied by Jorgensen, Seshadri & Whitmore (1991). For $p = 1/2$ the more well-known life distribution of Birnbaum & Saunders (1969) appears. Obviously $g \in \mathcal{B}_2$ since $f \in \mathcal{B}$. As $\varphi'$ is the mgf of $xf(x)$, the mgf of g is given by

$$\psi(s) = (p + q/\sqrt{1-c_4s})\cdot\varphi(s)\ .$$

Since the first factor corresponds to an MED, we see that $g \in \mathcal{T}_2$.

In fact, it is easy to prove that for all $f \in \mathcal{B}$, any pdf of the form $g(x) = C\cdot(1+cx)\cdot f(x)$ $(c \ge 0)$ belongs to $\mathcal{B}_2$ as well as $\mathcal{T}_2$. □

As $\mathcal{B} \subset \mathcal{T}$, one may hope that $\mathcal{B}_2 \subset \mathcal{T}_2$. Unfortunately, a counterexample shows that $\mathcal{B}_2 \not\subset \mathcal{T}_2$. Let $f(x) = Cx^{\beta-1}(1+x)^2e^{-x}$, $x > 0$ $(\beta > 0)$, which belongs to $\mathcal{B}_2$ since

$$f(x) = \lim_{\delta\to 0} C_\delta x^{\beta+1}\Big(\delta + \frac{1-\delta}{1 + x/\delta}\Big)^2(1+\delta x)^{-1/\delta}.$$

The mgf of f equals

$$\varphi(s) = C\cdot(1-s)^{-\beta}\left(1 + \frac{2\beta}{1-s} + \frac{\beta(\beta+1)}{(1-s)^2}\right).$$

Since $\varphi$ has zeros at $s = 1+\beta \pm i\sqrt{\beta}$, it follows that $f \notin \mathcal{T}_2$.

It can be shown that f is ID. It is unknown whether all pdf's in $\mathcal{B}_2$ are ID.

### 9.5 Extensions

The class $\mathcal{T}_2$ can be modified and extended in different ways. The class of $Po(\lambda)$–mixtures with $\lambda \sim \mathcal{T}_2$ is a natural discrete $\mathcal{T}_2$–class which contains $\mathcal{T}_d$ and all mixtures of geometric distributions. By looking at limit distributions for sums of the form $\Sigma c_kX_k$, with $c_k \in \mathbb{R}$ and the $X_k$'s independent and $X_k \sim$ MED, we get an extension of $\mathcal{T}_2$ to a class of distributions on $\mathbb{R}$ including $\mathcal{T}_e$. We now consider, slightly more carefully, another type of modification of $\mathcal{T}_2$.

Let $\mathcal{T}_r$, $r > 0$, be the class of probability distributions on $R_+ = [0,\infty)$ for which the mgf satisfies

$$\varphi'(s)/\varphi(s) = a + \int_{[0,\infty)}(t-s)^{-r}Q_r(dt)\ , \tag{9.5.1}$$

where $a \geq 0$ and $Q_r(dt)$ is a nonnegative measure on $\mathbb{R}_+$ satisfying, for $r \neq 1$,

$$\int_{[0,1)} t^{\min(0,1-r)} Q_r(dt) \leq \infty \quad \text{and} \quad \int_{[1,\infty)} t^{-r} Q_r(dt) < \infty$$

and without mass at 0 for $r \geq 1$. By an integration of (9.5.1) we get, for $r \neq 1$,

$$\varphi(s) = \exp\{as + \int_{[0,\infty)} \left((t-s)^{1-r} - t^{1-r}\right) \frac{1}{r-1} Q_r(dt)\} .$$

Obviously, for $r = 1$, $\mathcal{T}_r = \mathcal{T}$ and, for $r = 2$, $\mathcal{T}_r = \mathcal{T}_2$. Each class $\mathcal{T}_r$ is closed with respect to convolution and weak limits and $\mathcal{T}_{r_1} \subset \mathcal{T}_{r_2}$ for $r_1 < r_2$. The class of all ID distributions equals the weak closure of $\lim_{r\to\infty} \mathcal{T}_r$. An ID distribution on $\mathbb{R}_+$ belongs to $\mathcal{T}_r$ iff the Lévy measure has a density $\ell$ such that $y^{2-r}\ell(y)$ is CM. For $r < 1$, the class $\mathcal{T}_r$ equals the class of weak limits of finite convolutions of pdf's of the form $f(x) = C \cdot g_{1-r}(x) e^{-cx}$, $c \geq 0$, where $g_{1-r}$ is the pdf of a (strictly) Stable distribution of index $\alpha = 1-r$. Thus, for $r = 1/2$, inverse Gaussian distributions are convolved.

**Example 9.5.1** ***(Stable distribution).*** A Stable distribution of index $\alpha$ belongs to $\mathcal{T}_r$ for all $r \geq 1-\alpha$. This follows from the relation $(-s)^{\alpha-1} \propto \int (t-s)^{-r} t^{-\gamma} dt$, $\gamma = 2-\alpha-r$. □

**9.6 Notes.** The $\mathcal{T}_2$–distributions were introduced in B1981 for the purpose of finding a class of ID densities containing the MED's as well as the $\mathcal{B}$–densities. The attempt was not successful and I have often considered the class as a failure. Others have been much more interested in it and my name was even attached to it by Kent (1981). Unfortunately, some people have also called it B, e.g. Küchler & Lauritzen (1989). Earlier, Ito & McKean (1965, pp. 214–217) encountered the class in connection with first passage times for diffusion processes.

Theorem 9.1.1 – 5 and Example 9.2.1 are from B1981 whereas Theorems 9.1.6 and 9.1.7 are new. Theorem 9.1.5 can also essentially be found in Steutel's (1970) thesis.

The ID of the non–central $\chi^2$–distribution has been noted by several people, e.g. Ismail & Kelker (1979) and Hjort (1988). Example 9.2.3 is essentially taken from Shanbhag, Pestana & Sreehari (1977). Examples 9.2.4 and 9.2.5 are new. Example 9.2.6 is due to Berg & Forst (1982).

First passage time distributions have a rich literature. Theorem 9.3.1 dates essentially back to Karlin & McGregor (1957). Theorem 9.3.2 is due to Keilson (1981) who used it to verify that first passage time distributions are unimodal. Many earlier results can be found in Keilson (1979). It was noticed in the addendum to B1981 that Keilson's result makes it possible to characterize first passage time distributions. Yamazato (1988) gives the characterization problem a full treatment.

The results in Sections 9.4 and 9.5 stem essentially from B1981 and B1988a. Example 9.4.1 is taken from Jorgensen, Seshadri & Whitmore (1991).

# 10. SHOT–NOISE DISTRIBUTIONS

Some of the distributions encountered in Chapters 3–9 appear in a natural way as *shot–noise distributions*, i.e., as marginal distributions of shot–noise processes. This final short chapter is devoted to that topic.

In Section 10.1 basic facts about shot–noise processes are given. Moreover, it is verified that the Gamma and Negative Binomial distributions are shot–noise distributions. In Section 10.2 some classes of shot–noise distributions on $\mathbb{R}_+$ are studied. Discrete shot–noise distributions are treated in Section 10.3. Shot–noise distributions on $\mathbb{R}$ are not treated.

## 10.1 Shot–Noise Processes

In a general setting a shot–noise process is a process

$$X(\tau) = \sum_k Z(\tau,T_k),$$

where the $T_k$'s are points of a Poisson point–process on a space $\mathcal{T}$ and $\{Z(\cdot,t), t \in \mathcal{T}\}$ is a family of independent stochastic processes. The interpretation is that $Z(\tau,t)$ represents the random effect (noise) a possible event (shot) at $t$ has at $\tau$. The total effect is obtained by adding the effects of all events. The parameter $\tau$ may take values in $\mathcal{T}$ or in some other space. If $\tau$ and $t$ take values in $\mathbb{R}^m$ and if the distribution of $Z(\tau,t)$ depends on $\tau - t$ only, then $X(\cdot)$ is a stationary process.

The mgf of $X(\tau)$ admits a simple expression. In fact, for a Poisson process with intensity function $\lambda(t)$ it is given by

$$(10.1.1) \qquad E[\exp\{sX(\tau)\}] = \exp\left\{ \int_{\mathcal{T}} \lambda(t)\cdot(E[\exp\{sZ(\tau,t)\}] - 1)dt \right\}.$$

This formula, which has a simple multivariate extension, can be proved by conditioning on the number of events (points) in a compact region of $\mathcal{T}$. These points are independently distributed over the region according to a pdf proportional to $\lambda(t)$. The region is then expanded to cover in the limit the whole of $\mathcal{T}$.

A shot–noise distribution is ID since, for each $n$, the Poisson process can be considered to be the superposition of $n$ independent Poisson processes with intensities $\lambda(t)/n$.

Shot–noise processes have applications in many different fields. Some classical applications are given below.

*Current.* Let $Z(\tau,t)$ be the current at time $\tau$ induced by an electron that arrives at time $t$ at the anode in a vacuum tube. Letting $g$ be the response function, we then have $Z(\tau,t) = g(\tau-t)$ for $t \le \tau$ and zero elsewhere. The total current is given by $X(\tau) = \Sigma_{T_k \le \tau}\, g(\tau-T_k)$. This is the classical shot–noise process, which has a very large literature. □

*Insurance.* Let $X(\tau)$ be the discounted value at time $\tau$ of all future claims on an insurance company for damages occurring according to a Poisson process in time. Then $X(\tau) = \Sigma_{T_k > \tau}\, V_k \cdot \exp\{\rho\cdot(\tau-T_k)\}$, where $V_k$ is the claim (of random size) at time $T_k$. The $V_k$'s may be assumed to be iid. □

*Immigration and emigration.* Let individuals (or animals) arrive at a place according to a Poisson process. An individual arriving at time $t$ stays a random time $J^t$; all J–variables are assumed to be iid. The total number of individuals present at time $\tau$ is given by

$$X(\tau) = \sum_{T_k \le \tau} Z(\tau,T_k), \quad \text{where} \quad Z(\tau,t) = \mathbf{1}_{\{J^t > \tau - t\}}.$$

More generally, groups of individuals may arrive according to a Poisson process. The number of individuals arriving at $t$ is an rv $N^t$ and each individual stays a random time. Now

$$Z(\tau,t) = \sum_{i=1}^{N^t} \mathbf{1}_{\{J_i^t > \tau - t\}}. \qquad \square$$

There are many other examples. In the sequel we assume that the underlying Poisson process has constant intensity $\lambda$ and restrict attention to two particular stationary shot–noise processes, taking values in $\mathbb{R}_+$ and $\mathbb{N}_0$, respectively.

*Shot–noise process with values in* $\mathbb{R}_+$. Let

$$X(\tau) = \Sigma_{T_k \le \tau}\, V_k \cdot g(\tau - T_k),$$

where $g \ge 0$ and the variables $V_k$ are nonnegative and iid. According to (10.1.1) the mgf of $X(\tau)$ is given by

$$(10.1.2) \quad \varphi_{X(\tau)}(s) = \exp\{\lambda \int_{-\infty}^{\tau} \Big(E[\exp\{sV\cdot g(\tau-t)\}] - 1\Big)dt\} = \exp\{\lambda \int_0^{\infty} \Big(\varphi_V(sg(u)) - 1\Big)du\}.$$

***Shot–noise process with values in*** $\mathbb{N}_0$. Let

$$(10.1.3)\qquad X(\tau)=\Sigma_{T_k\le\tau}\,Z(\tau,T_k),\quad\text{with}\quad Z(\tau,t)=\sum_{i=1}^{N^t}1_{\{J_i^t>\tau-t\}}\,.$$

where all J– and N–variables are independent and iid within the J– and N–groups. By (10.1.1) with $\exp\{s\}=z$, the pgf of $X(\tau)$ is given by

$$(10.1.4)\qquad P_{X(\tau)}(z)\;=\;\exp\Big\{\,\lambda\int_0^\infty\big(P_N(z\overline{F}_J(u)+F_J(u))-1\big)du\Big\}\,,$$

where $P_N$ is the pgf of N and $F_J$ $(=1-\overline{F}_J)$ is the cdf of J.

**Example 10.1.1 (*Gamma distribution*).** Let $V\sim\text{Exp}(t_0)$ and $g(u)=e^{-\rho u}$, $u>0$. (This is the most natural response function.) Then, and by the substitution $t=t_0e^{\rho u}$,

$$\varphi_{X(\tau)}(s)=\exp\Big\{\lambda\int_0^\infty\Big(\frac{1}{1-\frac{s}{t_0}e^{-\rho u}}-1\Big)du\Big\}\;=\;\exp\Big\{\int_{t_0}^\infty\Big(\frac{1}{t-s}-\frac{1}{t}\Big)\frac{\lambda}{\rho}dt\Big\}\;=\;\Big(\frac{t_0}{t_0-s}\Big)^{-\lambda/\rho}.$$

Thus $X(\tau)$ is Gamma$(\lambda/\rho,\,t_0)$–distributed. We see that only the ratio $\lambda/\rho$ is of importance, which is quite natural. □

It can be shown that, for any point T in the Poisson process, $X(T)\sim$ Gamma$(\lambda/\rho+1,\,t_0)$, and, perhaps surprisingly, $X(T-)\sim$ Gamma$(\lambda/\rho,\,t_0)$. The last result can be verified to be equivalent to the result in Example 2.2.2 (Random reduction) though the formulations are quite different. If $\tau$ is an arbitrary time and now T is the *last point* before $\tau$ in the Poisson process, then as is well known, $\tau-T\sim\text{Exp}(\lambda)$. This may help the reader to see that $X(\tau)$ has the same distribution as $X(T-)$.

The next example is, by analogy, a dual of Example 2.2.3.

**Example 10.1.2 (*Negative Binomial distribution*).** Let N be geometrically distributed, i.e., $P_N(z)=\frac{p}{1-qz}$. Let further, $J\sim\text{Exp}(\rho)$, i.e., $\overline{F}_J(u)=e^{-\rho u}$. By (10.1.4),

$$P_{X(\tau)}(z)\;=\;\exp\Big\{\,\lambda\int_0^\infty\Big(\frac{p}{1-q\cdot(ze^{-\rho u}+1-e^{-\rho u})}-1\Big)du\Big\}\,.$$

Thus

$$P_{X(\tau)}(s+1)\;=\;\exp\Big\{\,\lambda\int_0^\infty\Big(\frac{1}{1-\frac{q}{p}s\cdot e^{-\rho u}}-1\Big)du\Big\}\;=\;\Big(1-\frac{q}{p}s\Big)^{-\lambda/\rho},$$

which is the mgf of a Gamma$(\lambda/\rho,\,p/q)$–distribution. Hence $X(\tau)\sim\text{NB}(\lambda/\rho,\,q)$. □

If the distribution of V in Example 10.1.1 is a finite mixture of Exponential distributions (MED), then $X(\tau)$ is a sum of independent Gamma variables. In fact, the points for which a specific Exponential distribution of the mixture is chosen also form a Poisson process but with reduced intensity. The Poisson processes corresponding to different Exponential distributions are independent.

## 10.2 Classes of Shot–Noise Distributions on $\mathbb{R}_+$

In this section we consider $X = X(0) = \Sigma_{k=1}^{\infty} V_k g(-T_k)$, where the $T_k$'s are points in a Poisson process of intensity $\lambda$ on $(-\infty, 0)$ We shall see what distributions are obtained for X when certain restrictions are put on $g \geq 0$ and the common distribution of the nonnegative rv's $V_k$. We assume that g is such that $X < \infty$ almost surely. (It can be shown that otherwise $X = \infty$ a.s.)

It is no restriction to assume that g is decreasing since g can be *rearranged* to be so. If g is piecewise constant, this is done by moving intervals on which g is large to the left and intervals on which g is small to the right. In the general case, the new g is defined to be an inverse of the right–continuous function $x \to \mu\{u;\, g(u) > x\}$, where $\mu$ is Lebesgue measure.

Since $X = X(0)$ is ID, $\varphi_X(s) = \exp\{\int_0^\infty (e^{sy}-1)L(dy)\}$, where $L(dy)$ is the Lévy measure. The left–extremity of the distribution of X must be zero. We also have

$$\varphi_X(s) = \exp\{\lambda \int_0^\infty (E[e^{sVg(u)}] - 1)du\} = \exp\{\lambda \int_0^\infty \Big(\int_0^\infty (e^{svg(u)} - 1)F_V(dv)\Big)\, du\}$$

Collecting contributions where $vg(u)$ is constant, we get, for $x > 0$,

$$\text{(10.2.1)} \qquad \overline{L}(x) = \int_{(x,\infty)} L(dy) = \lambda \int_0^\infty \overline{F}_V\Big(\frac{x}{g(u)}\Big)du = \lambda \int_{(0,\infty)} \overline{F}_V(x/y) g^{-1}(dy),$$

which is the general relation between L, $\overline{F}_V$ and g.

Changing the scale of u, we may without restriction assume that $\lambda = 1$. Now some different cases are studied.

A. ***The case $V \equiv 1$ and $\lambda = 1$.*** Since $\overline{F}_V(x)$ equals 1 for $x < 1$ and 0 for $x \geq 1$, (10.2.1) states that $\overline{L}(x) = g^{-1}(x)$ or $L(dx) = g^{-1}(dx)$. Certainly any nonnegative measure $L(dy)$ is of this form for some decreasing function g. Hence any ID distribution on $\mathbb{R}_+$ with left–extremity 0 is a shot–noise distribution. The self–decomposable (SD) distributions appear when g is logconvex. We have:

**Theorem 10.2.1.** *For a nonnegative and decreasing function* $g$, *the rv* $X=\Sigma_{k=1}^{\infty}g(-T_k)$ *is SD iff* $g$ *is logconvex and strictly decreasing.*

**Proof.** Assume that $X$ is SD. A distribution is SD iff $L$ has a density $\ell$ such that $x\ell(x)$ is decreasing. Let $b$ be the right–extremity of $\ell$. It easily follows that $\bar{L}(x)$ is strictly decreasing and continuous on $(0,b)$ and $\bar{L}(0+)=\infty$. Hence $g=\bar{L}^{-1}$ is strictly decreasing and continuous on $(0,\infty)$. Now $g(\bar{L}(x))=g(g^{-1}(x))=x$ for $x>0$. By a differentiation, $g'(\bar{L}(x))\cdot\ell(x)=-1$, and thus

$$g'(\bar{L}(x))/g(\bar{L}(x)) = -\frac{1}{x\ell(x)}\,.$$

We see that $g'/g$ is increasing on $(0,\infty)$, i.e., $g$ is *logconvex.*

The converse is rather obvious from the above. □

**Corollary.** *If* $g$ *is logconvex and strictly decreasing, then* $X=\Sigma_{k=1}^{\infty}V_kg(-T_k)$ *is SD for any distribution of the nonnegative iid rv's* $V_1, V_2, \ldots$

In fact, any distribution of $V$ can be obtained by mixing degenerate distributions. Mixing on the V–side corresponds to convolution (of convolution roots) on the X–side.

For $g(u)=c\cdot u^{-\beta}$, $\beta>1$, we get $\bar{L}(x)=g^{-1}(x) = c^{-1/\beta}x^{-1/\beta}$ and hence $\ell(x)=c'\cdot x^{-1-1/\beta}$, corresponding to a Stable distribution with index $\alpha=1/\beta$. By the mixing reasoning it follows that the distribution is Stable also when there are random V–effects.

B. ***Exponentially distributed V's.*** Let $\lambda=1$ and $V_k\sim \mathrm{Exp}(1)$. In this case

$$\bar{L}(x) = \int_0^{\infty}\bar{F}_V(x/g(u))du = \int_0^{\infty}\exp\{-x/g(u)\}du\,.$$

We see that $\bar{L}$ is completely monotone (CM) and hence so is $\ell$; i.e., $X\sim\mathcal{T}_2$; cf. Chapter 9. Any $\mathcal{T}_2$–distribution with left–extremity zero can be obtained in this way. For a $\mathcal{T}_2$–distribution we have $\ell(x)=\int e^{-xt}Q(dt)$. Thus it suffices to choose $g$ such that $t^{-1}Q(dt)=h^{-1}(dt)$, where $h(u)=1/g(u)=t$. In particular the GGC's can be obtained. Moreover:

**Theorem 10.2.2.** *For a nonnegative and decreasing function* $g$ *and iid rv's* $V_k\sim Exp(1)$, *the distribution of the rv* $X=\Sigma_{k=1}^{\infty}V_kg(-T_k)$ *is a GGC iff* $g$ *is logconvex and strictly decreasing.*

**Proof.** Assume that $X \sim \mathcal{T}$. For a GGC, $Q(dt) = q(t)dt$, where $q(t) = U(t)$. Hence $h^{-1}(t) = \int_{(0,t]} \tilde{t}^{-1}U(\tilde{t})d\tilde{t}$ is strictly increasing and continuous on some interval $(b,\infty)$. It follows that $g = 1/h$ is strictly decreasing and continuous on $(0, \infty)$. Further,

$$u = \int_{(0,\, 1/g(u)]} \tilde{t}^{-1}U(\tilde{t})d\tilde{t}\ .$$

By differentiation we get

$$1 = -\frac{U(1/g(u))}{1/g(u)} \cdot \frac{g'(u)}{(g(u))^2},$$

i.e.,

$$g'(u)/g(u) = -1/U(1/g(u)),$$

which is an increasing function. Hence $g$ is logconvex. The "only if" part is proved. Moreover, by an integration we get

$$g(u) = \exp\{-W^{-1}(u)\}, \quad \text{where} \quad W(x) = \int_{-\infty}^{x} U(e^y)dy\ .$$

Any logconvex and strictly decreasing function $g$ can be represented in this way by an appropriate choice of $U$. The "if" part follows. □

By the mixing reasoning, $X = \Sigma_{k=1}^{\infty} V_k g(-T_k) \sim \mathcal{T}_2$ also when $V_k \sim \text{MED}$ for general $g$. If, moreover, $g$ is logconvex and strictly decreasing, then $X \sim \mathcal{T}$.

C. ***Exponential response function.*** Instead of specifying the distribution of the V's, we now specify $g(u)$ to be $e^{-\rho u}$, $\rho > 0$. Again it is no restriction to assume that $\lambda = 1$. Now, since $g$ is logconvex, $X$ is certainly SD. In fact, that was seen already in Section 2.3. However, all SD distributions with left–extremity zero cannot be obtained in this way. From the first part of (10.2.1) we get, by substituting $y = x/g(u) = xe^{\rho u}$,

$$\underline{L}(x) = \int_x^{\infty} \overline{F}_V(y)(\rho y)^{-1}dy$$

and hence

$$\rho x\ell(x) = \overline{F}_V(x).$$

Thus precisely the SD distributions for which $r(x) = x\ell(x)$ has a finite limit as $x \downarrow 0$ are shot–noise distributions of the considered type.

Moreover, *any GGC with* $\int U(dt) < \infty$ *is such a shot–noise distribution.* The distribution of $V$ and $\rho$ are chosen so that $\overline{F}_V(x) = \rho x\ell(x) = \rho \int e^{-xt}U(dt)$. The U–measure, appropriately normalized, is a mixing measure and $V \sim \text{MED}$.

The above is summarized in the following table.

**Table 10.2.1.** *Classes of shot–noise distributions corresponding to different classes of response distributions and response functions. (The shot–noise distributions all have left–extremity zero.)*

| Class of response functions | $V = 1$ | Class of distributions for $V \geq 0$: full class | Exp | MED |
|---|---|---|---|---|
| full class | ID | ID | $\mathcal{T}_2$ | $\mathcal{T}_2$ |
| {logconvex and ↓} | SD | SD | $\mathcal{T}$ | $\mathcal{T}$ |
| {exponential} | No name | SD; $r(0+)<\infty$ | Gamma | $\mathcal{T}$ ; $\int U(dt) < \infty$ |

## 10.3. Discrete Shot–Noise Distributions

Every ID distributions on $\mathbb{N}_0$ can be obtained as the distribution of $X = X(0) = \Sigma_{k=1}^{\infty} g(-T_k)$. The function $g$ with values in $\mathbb{N}_0$ is just chosen so that $L(\{j\}) = \mu\{u;\, g(u) = j\}$; $L$ is the Lévy measure. However, the restriction to be imposed on $g$ to get the class of discrete SD distributions, for instance, is rather artificial. We turn to shot–noise processes of the form (10.1.3).

For $\lambda = 1$ the pgf of $X = X(0)$ is given by

$$P_X(z) = \exp\Big\{\int_0^{\infty} \big(P_N(z\overline{F}_J(u) + F_J(u)) - 1\big)du\Big\}.$$

It is not possible to get all discrete ID distributions as shot–noise distributions in this case. For example, for $N \equiv 1$, only the Poisson distribution appears and nothing else. Two cases, corresponding to (B) and (C) in Section 10.2, are particularly interesting.

B. ***Geometrically distributed N.*** When $N$ has a Geometric distribution, we have

$$P_X(z) = \exp\Big\{\int_0^{\infty} \Big(\frac{p}{1 - q(z\overline{F}_J(u) + F_J(u))} - 1\Big)du\Big\},$$

and hence

$$P_X(s+1) = \exp\Big\{\int_0^{\infty} \Big(\frac{1}{1 - s\frac{q}{p}\overline{F}(u)} - 1\Big)du\Big\}.$$

Thus the distribution of $X$ is a $Po(\tilde{\lambda})$–mixture with $\tilde{\lambda} = \Sigma_{k=1}^{\infty} V_k g(-T_k)$, where

$g(u) = (q/p)\overline{F}(u)$ and $V_k \sim \mathrm{Exp}(1)$; cf. (10.1.2). This is a discrete $\mathcal{T}_2$–distribution but with a restriction on the corresponding Q–measure since $g(0)$ is finite; the Q–measure is zero on the interval $(0, p/q)$. When $g$ is logconvex, i.e., when $J \sim$ DFR (decreasing failure rate), a GNBC is obtained but the U–measure of the corresponding GGC has no mass in $(0, p/q)$.

Discrete $\mathcal{T}_2$–distributions are also obtained when the distribution of N is a mixture of Geometric distributions. GNBC's are obtained if, moreover, $J \sim$ DFR.

C. ***Exponential distribution for J.*** When J has an Exponential distribution, discrete SD distributions are obtained, though not all. Because of the lack of memory of the Exponential distribution, the process $X(\tau)$ can be decomposed as $X(\tau+h) = \Upsilon_h \circ X(\tau)$ + innovation, where the operator $\Upsilon_h$ corresponds to binomial thinning; cf. the original definition of discrete SD in Section 2.3. When the distribution of N is a mixture of Geometric distributions, GNBC's are obtained. Any GNBC with finite total spectral mass is possible.

**10.4 Notes.** Shot–noise processes have a very large literature. Parzen (1962) provides a particularly good exposition. A 1983 survey on shot–noise processes and shot–noise distributions is given in B1988b.

The result that the Gamma distribution is a shot–noise distribution dates at least back to Bartlett (1957). A later paper on the result is that by Lawrance (1982). The fact that also the Negative Binomial distribution can be viewed as a shot–noise distribution does not seem to be an old result. It is mentioned in B1982, 1988b and in McKenzie (1987). Related results concerning spatial quadrat counts are given by e.g. Diggle & Milne (1983) and Gregoire (1983).

In part the results in Sections 10.2 are from B1982. However, Theorems 10.2.1 and 10.2.2 are new. The self–decomposability when the response function is exponential has been treated by several authors, e.g. Vervaat (1979) and Wolfe (1982). The results in Section 10.3 are also in part new. McKenzie (1986, 1987) has related results for autoregressions. Other related results are given by van Harn & Steutel (1991).

# FINAL WORDS

**Future.** Without any doubt, the main novelty in this work is the systematic treatment of hyperbolically completely monotone densities. The author believes that the work is not the end of the HCM–story. Ever since they were introduced, the properties of logconvexity, complete monotonicity, logconcavity, and total positivity have found a large number of applications in many different fields. I conjecture that the same will happen with the HCM–property and that this work is the end of the beginning of the HCM–story. It is to be hoped that, either in the distant future or sometime soon, a probability mechanism generating the full class of HCM–densities will appear.

**Apology.** Some years ago the author heard a voice saying "*Though infinite divisibility is interesting, it is not very interesting to find out whether or not particular distributions are ID*". The statement has its roots in the fact that there are few applications of the knowledge of the ID of a particular distribution. Contrary to this, the author's position has long been roughly as follows: "*Infinite divisibility is not a very important property, but it is interesting to find out whether or not particular distributions are ID*". However curious and illogical this statement may seem to be, it is motivated as follows:

Since the class of ID distributions is extremely large, it is not a very interesting class. In fact, as the research during the last two decades has shown, ID seems to be more a rule than an exception. This is not surprising if one considers that the class of (univariate) distributions which are ID with respect to the maximum operation contains all distributions. On the other hand, to investigate whether or not ID holds for a particular distribution may lead to a deep insight into the structure of that and other distributions and also to a lot of by–products. (Cf. Riemann's hypothesis in mathematics, for example.) This work would certainly not have been written had not Steutel (1973) asked whether the Lognormal distribution is ID and had not Thorin (1977a,b) attacked and solved this problem.

# BIBLIOGRAPHY

The bibliography is not exhaustive. However, it also contains some papers which are not quoted in the text but have content relevant for the field.


Abramowitz, M. & Stegun, I.A. (1964). *Handbook of Mathematical Functions*. U.S. Department of Commerce, Washington.

Bain, L. (1983). Gamma distribution. In *Encyclopedia of Statistical Sciences*, Vol 3. Eds: N.L. Johnson & S. Kotz. Wiley, New York.

Barlow, R.E. & Proschan, F. (1965). *Mathematical Theory of Reliability*. Wiley, New York.

Barlow, R.E. & Proschan, F. (1975). *Statistical Theory of Reliability and Life Testing: Probability Models*. Holt, Rinehart and Winston, New York.

Barndorff–Nielsen, O. & Halgreen, C. (1977). Infinite divisibility of the hyperbolic and generalized inverse Gaussian distribution. *Z. Wahrsch. verw. Gebiete* **38**, 309–312.

Barndorff–Nielsen, O., Blaesild, P., & Halgreen, C. (1978). First hitting time models for the generalized inverse Gaussian distribution. *Stochastic Processes Appl.* **7**, 49–54.

Barndorff–Nielsen, O., Kent, J.T., & Sorensen, M. (1982). Normal variance–mean mixtures and z–distributions. *Intern. Statist. Review* **50**, 145–159.

Bartlett, M.S. (1957). Contribution to discussion of papers by Dr. Gani and Mr. Kendall. *J.R. Statist. Soc.*, B **19**, 220–21.

Berg, C. (1981). The Pareto distribution is a generalized Γ–convolution – a new proof. *Scand. Actuarial J.* **1981**, 117–119.

Berg, C. & Forst, G. (1979). Infinitely divisible probability measures and potential kernels. In *Lecture Notes in Mathematics* **706**, pp 22–35. Springer–Verlag, Berlin–Heidelberg–New York.

Berg, C. & Forst, G. (1980). Uendeligt delbare mål (Infinitely divisible measures). Lecture notes, Dept. of Math., Univ. of Copenhagen. (In Danish.)

Berg, C. & Forst, G. (1982). A convolution equation relating to generalized Γ–convolutions and the Bondesson class. *Scand. Actuarial J.* **1982**, 171–175.

Berg, C. & Forst, G. (1983). Multiply self–decomposable probability measures on $R_+$ and $Z_+$. *Z. Wahrsch. verw. Gebiete* **62**, 147–163.

Bingham, N., Goldie, C. & Teugels, J. (1987). *Regular Variation*. Cambridge University Press, London.

Birnbaum, Z.W. & Saunders, S.C. (1969). A new family of life distributions. *J. Appl. Probab.* **6**, 319–327.

Boas, R.P. (1981). *A Primer of Real Functions*. The Math. Assoc. of America.

Bondesson, L. (1978). On infinite divisibility of powers of a Gamma variable. *Scand. Actuarial J.* **1978**, 48–61.

Bondesson, L. (1979a). A general result on infinite divisibility. *Ann. Probab.* **7**, 965–979.

Bondesson, L. (1979b). On the infinite divisibility of products of powers of Gamma variables. *Z. Wahrsch. verw. Gebiete* **49**, 171–175.

Bondesson, L. (1979c). On generalized gamma and generalized negative binomial convolutions, Parts I & II. *Scand. Actuarial J.* **1979**, 125–166.

Bondesson, L. (1981). Classes of infinitely divisible distributions and densities. *Z. Wahrsch. verw. Gebiete* **57**, 39–71. (Correction and Addendum, ibid. **59**, 277.)

Bondesson, L. (1982). On simulation from infinitely divisible distributions. *Adv. Appl. Probab.* 14, 855–869

Bondesson, L. (1984). New results on generalized Gamma convolutions and the B–class. *Scand. Actuarial J.* **1984,** 197–209.

Bondesson, L. (1987). On the infinite divisibility of the half–Cauchy and other decreasing densities and probability functions on the non–negative line. *Scand. Actuarial J.* **1987,** 225–247.

Bondesson, L. (1988a). A remarkable property of generalized Gamma convolutions. *Probab. Th. Rel. Fields* **78**, 321–333.

Bondesson, L. (1988b). Shot–noise processes and shot–noise distributions. In *Encyclopedia of Statistical Sciences*, Vol 8. Eds: N.L. Johnson & S. Kotz. Wiley, New York.

Bondesson, L. (1988c). The $T_1$– and $T_2$–classes of distributions. In *Encyclopedia of Statistical Sciences*, Vol 9. Eds: N.L. Johnson & S. Kotz. Wiley, New York.

Bondesson, L. (1990). Generalized gamma convolutions and complete monotonicity. *Probab. Th. Rel. Fields* **85**, 181–194.

Bondesson, L. (1991). A characterization of first passage time distributions for random walks. *Stochastic Processes Appl.* **39**, 81–88.

Boswell, M.T. & Patil, G.P. (1970). Chance mechanisms generating the negative binomial distribution. In *Random Counts in Scientific Work*, Vol I. Ed.: G.P. Patil. Pennsylvania State University Press, University Park.

Chang, D.K. (1989). On infinitely divisible discrete distributions. *Utilitas Mathematica* **36**, 215–217.

Clausius, R. (1858). Über die mittlere Länge der Wege .. *Annalen der Physik* **105**, 239–258.

Cox, D.R. (1981). Statistical analysis of time series: Some recent developments. *Scand. J. Statist.* **8**, 93–115.

Dharmadhikari, S. & Joag–Dev, K. (1988). *Unimodality, Convexity, and Applications.* Academic Press, New York.

Diggle, P.J. & Milne, R.K. (1983). Negative binomial quadrat counts and point processes. *Scand. J. Statist.* **10**, 257–316.

Donoghue, W. (1974). *Monotone Matrix Functions and Analytic Continuation.* Springer–Verlag, Berlin–Heidelberg–New York.

Ebbesen, G. (1983). Generaliserede Γ–foldninger og Bondesson klassen. Dept. of Mathematics, University of Copenhagen. Master's thesis (in Danish).

Edwards, H.M. (1974). *Riemann's Zeta Function.* Academic Press, New York and London.

Epstein, B. (1977). Infinite divisibility of Student's t–distribution. *Sankhya* B**39**, 103–120.

Erdélyi, A. et al. (1954). *Tables of Integral Transforms, Vol I.* McGraw Hill, New York.

Feller, W. (1966). Infinitely divisible distributions and Bessel functions associated with random walks. *SIAM J. Appl. Math.* **14**, 864–875.

Feller, W. (1971). *An Introduction to Probability Theory and Its Applications,* Vol II, 2nd ed. Wiley, New York.

Ferguson, T. (1962). Location and scale parameters in exponential families of distributions. *Ann. Math. Statist.* **33**, 986–1001.

Forst, G. (1979). A characterization of self–decomposable probabilities on the half–line. *Z. Wahrsch. verw. Gebiete* **49,** 349–352.

Forst, G. (1981). Subordinates of the Poisson semigroup. *Z. Wahrsch. verw. Gebiete* **55**, 35–40.

Forst, G. (1984). Self–decomposability on R and Z. In *Probability Measures on Groups VII,* 99–115. Lecture Notes in Math., 1064, Springer–Verlag, Berlin–Heidelberg–New York.

Goldie, C.M. (1967). A class of infinitely divisible distributions. *Proc. Camb. Phil. Soc.* **63,** 1141–1143.

Goovaerts, M.J., D'Hooge, L. & de Pril, N. (1977a). On a class of generalized Γ—convolutions (part I). *Scand. Actuarial J.* **1977,** 21—30.

Goovaerts, M.J., D'Hooge, L. & de Pril, N. (1977b). On the infinite divisibility of the product of two Γ—distributed stochastic variables. *Appl. Math. Comput.* **3**, 127—135.

Goovaerts, M.J., D'Hooge, L. & de Pril, N. (1978). On the infinite divisibility of the ratio of two Gamma distributed random variables. *Stochastic Processes Appl.* **7**, 291—298.

Grandell, J. (1985). Stochastic Models of Air Pollutant Concentrations. *Lecture Notes in Statistics* **30**. Springer—Verlag, Berlin—Heidelberg—New York.

Greenwood, M. & Yule, G.U. (1920). An inquiry into the nature of frequency distributions representative of multiple happenings with particular reference to the occurence of multiple attacks of disease or of repeated accidents. *J. R. Statist. Soc.*, **83**, 255—279.

Gregoire, G. (1983). Negative binomial distributions for point processes. *Stochastic Processes Appl.* **16**, 179—188.

Grosswald, E. (1976). The Student t—distribution of any degree of freedom is infinitely divisible. *Z. Wahrsch. verw. Gebiete* **36**, 103—109.

Halgreen, C. (1979). Self—decomposability of the generalized inverse Gaussian and hyperbolic distributions. *Z. Wahrsch. verw. Gebiete* **47**, 13—18.

Hall, P. (1981). A comedy of errors: The canonical form for a stable characteristic function. *Bull. London Math. Soc.* **13**, 23—27.

Hammersley, J.M. (1960). On the statistical loss of long—period comets from the solar system II. In *Proc. 4th Berkeley Symp. on Math. Stat. and Probability,* Vol III.

Hansen, B.G. (1988) *Monotonicity Properties of Infinitely Divisible Distributions.* CWI—tract 69, Centre for mathematics and computer science, Amsterdam.

van Harn, K. (1978). *Classifying Infinitely Divisible Distributions by Functional Equations.* Mathematical Centre Tracts **103**, Matematisch Centrum, Amsterdam.

van Harn, K. & Steutel, F.W. (1991). Stability equations for processes with stationary independent increments using branching processes and Poisson mixtures. To appear.

Herglotz, G. (1911). Über Potenzreihen mit positivem, reelem Teil in Einheitskreis. *Leipzig Bericht* **63**, 501—511.

Hjort, N.L. (1988). The eccentric part of the non—central chi—square. *Amer. Statist.* **42**, 130—132.

Hougaard, P. (1986). Survival models for heterogeneous populations derived from stable distributions. *Biometrika* **73**, 387—396.

Ibragimov, I.A. (1956). On the composition of unimodal distributions. *Theor. Probab. Appl.* **1**, 255—260.

Irwin, J.O. (1975). The generalized Waring distribution, Parts I—III. *J. R. Statist. Soc.*, A **138**, 18—31, 204—227, 374—384.

Ismail, M.E.H. (1977). Bessel functions and the infinite divisibility of the Student t—distribution. *Ann. Probab.* **5**, 582—585.

Ismail, M.E.H. & Kelker, D. (1976). The Bessel polynomials and the Student t—distribution. SIAM *J. Math. Anal.* **7**, 82—91.

Ismail, M.E.H. & Kelker, D. (1979). Special functions, Stieltjes transforms and infinite divisibility. *SIAM J. Math. Anal.* **10**, 884—901.

Ito, K. & McKean, H.P. (1965). *Diffusion Processes and Their Sample Paths.* Springer—Verlag, Berlin.

Johnson, N.L. & Kotz, S. (1969). *Distributions in Statistics, Discrete Distributions.* Wiley, New York.

Johnson, N.L. & Kotz, S. (1970a). *Distributions in Statistics, Continuous Univariate Distributions —1.* Wiley, New York.

Johnson, N.L. & Kotz, S. (1970b). *Distributions in Statistics – Continuous Univariate Distributions –2*. Wiley, New York.

Jorgensen, B. (1981). Statistical properties of the generalized inverse Gaussian distribution. *Lecture Notes in Statistics* **9**. Springer–Verlag, Berlin – Heidelberg – New York.

Jorgensen, B., Seshadri, V., & Whitmore, G.A. (1991). On the mixture of the inverse Gaussian distribution with its complementary reciprocal. *Scand. J. Statist.* **18**, 77–89.

Jurek, Z.J. & Vervaat, W. (1983). An integral representation for Banach space valued random variables. *Z. Wahrsch. verw. Gebiete* **62**, 247–262.

Kallenberg, O. (1976). *Random Measures.* Akademie Verlag, Berlin.

Karlin, S. (1968). *Total Positivity, Vol I.* Stanford Univ. Press, Stanford.

Karlin, S. & McGregor, J.L. (1957). The differential equations of birth–and–death processes, and the Stieltjes moment problem. *Trans. Amer. Math. Soc.* **85**, 489–546.

Katti, S.K. (1979). Infinite divisibility of discrete distributions III. In *Analytic Function Methods in Probability Theory*, 165–171. Colloquia Mathematica Societa Janos Bolyai 21, North Holland, Amsterdam.

Keilson, J. (1979). Markov Chain Models – Rarity and Exponentiality. Springer–Verlag, Berlin–Heidelberg–New York.

Keilson, J. (1981). On the unimodality of passage time densities in birth–death processes, *Statistica Neerlandica* **25**, 49–55.

Kelker, D. (1971). Infinite divisibility and variance mixtures of the normal distribution. *Ann. Math. Statist.* **42**, 802–808.

Kendall, D.G. & Harding, E.F. (1973). *Stochastic Analysis.* Wiley, New York.

Kent, J. (1978). Some probabilistic properties of Bessel functions. *Ann. Probab.* **6**, 760–770.

Kent, J. (1980). Eigenvalue expansions for diffusion hitting times. *Z. Wahrsch. verw. Gebiete* **52**, 309–319.

Kent, J.T. (1981). Convolution mixtures of infinitely divisible distributions. *Math. Proc. Camb. Phil. Soc.* **90**, 141–153.

Kent, J.T. (1982). The spectral decomposition of a diffusion hitting time. *Ann. Probab.* **10**, 207–219.

Kent, J.T. (1983). On the appearance of a multivariate exponential distribution in sojourn times for birth–death and diffusion processes. In *Probability, Statistics and Analysis.* Eds: J.F.C. Kingman & G.E.H. Reuter. Cambridge University Press, London, pp. 161–175.

Kent, J.T. & Longford, N.T. (1983). An eigenvalue decomposition for first hitting time distributions in random walks. *Z. Wahrsch. verw. Gebiete* **63**, 71–84.

Kingman, J.F.C. (1972). *Regenerative Phenomena.* Wiley, London.

Krein, M.G. & Nudelman, A.A. (1977). *The Markov Moment Problem and Extremal Problems.* Transl of Math. Monographs, **50**. Amer. Math. Soc., Providence, Rhode Island.

Kristiansen, G.K. (1991). A proof of Steutel's conjecture. Manuscript.

Kumar, A. & Schreiber, B.M. (1978). Characterization of subclasses of class L probability distributions. *Ann. Probab.* **6**, 279–293.

Küchler, U. & Lauritzen, S.L. (1989). Exponential families, extreme point models and minimal space–time invariant functions for stochastic processes with stationary independent. *Scand. J. Statist.* **16**, 237–261.

Lai, C.D. & Moore, T. (1984). Logistic distribution in generalization. Manuscript.

Lawrance, A.J. (1982). The innovation distribution of gamma distributed autoregressive process. *Scand. J. Statist.* **9**, 234–236.

Loeve, M. (1978). Calcul des probabilités. Chapter 12 in *Abregé d'histoire des Mathématiques II.* Ed.: J. Dieudonné. Hermann, Paris.

Lukacs, E. (1970). *Characteristic Functions,* 2nd ed. Griffin, London.

Lukacs, E. (1983). *Developments in Characteristic Function Theory.* Griffin, London

Maloshevskii, S.G. (1972). Infinite divisibility of a certain family of functions. *Teor. Funktsii Funktsional. Analis Priloz.* **16**, 212–214.

Marshall, A.W. & Olkin, I. (1979). *Inequalities: Theory of Majorization and Its Applications.* Academic Press, New York.

Masuda, Y. (1988). First passage times of birth–death processes and simple random walks, *Stochastic Process. Appl.* **29**, 51–63.

McKenzie, E. (1986). Autoregressive moving–average processes with negative–binomial and geometric marginal distributions. *Adv. Appl. Prob.* **18**, 679–705

McKenzie, E. (1987). Innovation distributions for Gamma and negative binomial autoregressions. *Scand. J. Statist.* **14,** 79–85.

Ostrovskii, I.V. (1985). On an infinitely divisible factorization. *Selected Transl. in Math. Statist. and Probability,* Vol **16**. Amer. Math. Soc., Providence, Rhode Island.

Parzen, E. (1962). *Stochastic Processes.* Holden–Day, San Francisco.

Pearson, K. (1895). Contribution to the mathematical theory of evolution, II: skew variation. *Phil. Trans. R. Soc. London* (A) **186**, 343–414.

Petrov, V.M. (1975). *Sums of Independent Random Variables.* Springer–Verlag, Berlin–Heidelberg–New York.

Pitman, J.W. & Yor, M. (1981). Bessel processes and infinitely divisible laws. In *Stochastic Integrals (D. Williams, ed). Lecture Notes in Math.* **851**, 285–370. Springer–Verlag, Berlin–Heidelberg–New York.

Quenouille, M.H. (1949). A relation between the logarithmic, Poisson, and negative binomial series. *Biometrics* 5, 162–164.

Rohatgi, V.K., Steutel, F.W., & Szekely, G.J. (1990). Infinite divisibility of products and quotients of iid random variables. *Math. Scientist* **15**, 53–59.

Röhme, D. & Thorburn, D. (1981). Quantitative cell fusion: Derivation and application of theoretical models. *Somatic Cell Genetics* **7**, 43–57.

Ruben, H. (1960). On the distribution of the weighted difference of two independent Student variables. *J. R.Statist. Soc.* B **22**, 188–194.

Ruegg, A.F. (1970). A characterization of certain infinitely divisible laws. *Ann. Math. Statist.* **41**, 1354–1356.

Ruzsa, I.Z. & Szekely, G.J. (1988). *Algebraic Probability Theory.* Wiley, New York.

Schoenberg, I.J. (1951). On Polya frequency functions I; the totally positive functions and their Laplace transforms. *J. Anal. Math. Jerusalem* 1, 331–374.

Scott, W.T. (1968). Analytic studies of cloud droplet coalescence I. *J. Athmospheric Sciences* **25**, 54–65.

Shanbhag, D.N. & Sreehari, M. (1977). On certain self–decomposable distributions. *Z. Wahrsch. verw. Gebiete* **38**, 217–222.

Shanbhag, D.N. & Sreehari, M. (1979). An extension of Goldie's result and further results in infinite divisibility. *Z. Wahrsch. verw. Gebiete* **47**, 19–26.

Shanbhag, D.N., Pestana, D., & Sreehari, M. (1977). Some further results in infinite divisibility. Proc. Camb. Phil. Soc. 82, 289–295.

Springer, M.D. (1979). *The Algebra of Random Variables.* Wiley, New York.

Steutel, F.W. (1967). Note on infinite divisibility of exponential mixtures. *Ann. Math. Statist.* **38**, 1303–1305.

Steutel, F.W. (1968). A class of infinitely divisible mixtures. *Ann. Math. Statist.* **39**, 1153–1157.

Steutel, F.W. (1969). Note on completely monotone densities. *Ann. Math. Statist.* **40**, 1130–1131.

Steutel, F.W. (1970). Preservation of infinite divisibility under mixing and related topics. *Mathematical Center Tracts* **33**, Matematisch Centrum, Amsterdam.

Steutel, F.W. (1971). On the zeros of infinitely divisible densities. *Ann. Math. Statist.* **42**, 812–815.

Steutel, F.W. (1973). Some recent results in infinite divisibility. *Stochastic Processes Appl.* **1**, 125–143.

Steutel, F.W. (1979). Infinite divisibility in theory and practice. *Scand. J. Statist.* **6**, 57–62.

Steutel, F.W. & van Harn, K. (1979). Discrete analogues of self–decomposability and stability. *Ann. Probab.* **7**, 893–899.

Stigler, S.M. (1986). *The History of Statistics: The Measurement of uncertainity before 1900.* Belknap Press, Harvard.

Stoyanov, J. (1987). *Counterexamples in Probability.* Wiley, New York.

Stuart, A. (1962). Gamma distributed products of random variables. *Biometrika* **49**, 564–565.

Sumita, U. & Masuda, Y. (1985). On first passage time structure of random walks. *Stochastic Processes. Appl.* **20**, 133–147.

Takano, K. (1989b). On mixtures of the normal distribution by the generalized gamma convolutions. *Bull. Fac. Sci. Ibaraki Univ.* **21**, 29–41. (Correction and Addendum, ibid **22**.)

Teugels, J.L. (1971). Probability density functions which are their own characteristic functions. Bull. Soc. Math. Belgique, **23**, 1971

Thorin, O. (1977a). On the infinite divisibility of the Pareto distribution. *Scand. Actuarial J.* **1977**, 31–40.

Thorin, O. (1977b). On the infinite divisibility of the lognormal distribution. *Scand. Actuarial J.* **1977**, 121–148.

Thorin, O. (1978a). An extension of the notion of a generalized Γ–convolution. *Scand. Actuarial J.* **1978**, 141–149.

Thorin, O. (1978b). Proof of a conjecture of L. Bondesson concerning infinite divisibility of powers of a gamma variable. *Scand. Actuarial J.* **1978**, 151–164.

Tripathi, R.C. (1985). Negative binomial distribution. In *Encyclopedia of Statistical Sciences*, Vol 6. Eds: N.L. Johnson & S. Kotz. Wiley, New York.

Urbanik, K. (1972). Limit laws for sequences of normed sums satisfying some stability conditions. In *Multivariate Analysis* III. Ed.: P.R. Krishnaiah, 225–237. Academic Press, New York.

Vervaat, W. (1979). On a stochastic difference equation and a representation of positive infinitely divisible random variables. *Adv. in Appl. Probab.* **11**, 750–783.

Watson, E.J. (1981). *Laplace Transforms.* Van Nostrand, New York.

Widder, D. (1946). *The Laplace Transform.* Princeton Univ. Press.

Williams, E.J. (1977). Some representations of stable random variables as products. *Biometrika* **64**, 167–169.

Wolfe, S.J. (1982). On a continuous analogue of the stochastic difference equation $X_n = \rho X_{n-1} + B_n$. *Stochastic Processes Appl.* **12**, 301–312.

Xekalaki, E. (1983). Infinite divisibility, completeness and regression properties of the univariate generalized Waring distribution. *Ann. Inst. Statist. Math.* **35**, 279–283.

Xekalaki, E. & Panaretos, J. (1988). On the association of the Pareto and the Yule distributions. *Th. Probab. Appl.* **33**, 191–195.

Yamazato, M. (1978). Unimodality of infinitely divisible distribution functions of class L. *Ann. Probab.* **4**, 523–531.

Yamazato, M. (1988). Characterization of the class of upward first passage time distributions of birth and death processes and related results. J. Math. Soc. Japan, **40**, 477–499.

Yamazato, M. (1990). Hitting time distributions of single points for 1–dimensional generalized diffusion processes. *Nagoya Math. J.* **119**, 143–172.

Zolotarev, V.M. (1986). *One–dimensional Stable Distributions.* Translations of mathematical monographs, Vol 65. Amer. Math. Soc., Providence, Rhode Island.

# AUTHOR INDEX

# SUBJECT INDEX

Vol. 1: R.A. Fisher: An Appreciation. Edited by S.E. Fienberg and D.V. Hinkley. XI, 208 pages, 1980.

Vol. 2: Mathematical Statistics and Probability Theory. Proceedings 1978. Edited by W. Klonecki, A. Kozek, and J. Rosinski. XXIV, 373 pages, 1980.

Vol. 3: B.D. Spencer, Benefit-Cost Analysis of Data Used to Allocate Funds. VIII, 296 pages, 1980.

Vol. 4: E.A. van Doorn, Stochastic Monotonicity and Queueing Applications of Birth-Death Processes. VI, 118 pages, 1981.

Vol. 5: T. Rolski, Stationary Random Processes Associated with Point Processes. VI, 139 pages, 1981.

Vol. 6: S.S. Gupta and D.-Y. Huang, Multiple Statistical Decision Theory: Recent Developments. VIII, 104 pages, 1981.

Vol. 7: M. Akahira and K. Takeuchi, Asymptotic Efficiency of Statistical Estimators. VIII, 242 pages, 1981.

Vol. 8: The First Pannonian Symposium on Mathematical Statistics. Edited by P. Révész, L. Schmetterer, and V.M. Zolotarev. VI, 308 pages, 1981.

Vol. 9: B. Jørgensen, Statistical Properties of the Generalized Inverse Gaussian Distribution. VI, 188 pages, 1981.

Vol. 10: A.A. McIntosh, Fitting Linear Models: An Application of Conjugate Gradient Algorithms. VI, 200 pages, 1982.

Vol. 11: D.F. Nicholls and B.G. Quinn, Random Coefficient Autoregressive Models: An Introduction. V, 154 pages, 1982.

Vol. 12: M. Jacobsen, Statistical Analysis of Counting Processes. VII, 226 pages, 1982.

Vol. 13: J. Pfanzagl (with the assistance of W. Wefelmeyer), Contributions to a General Asymptotic Statistical Theory. VII, 315 pages, 1982.

Vol. 14: GLIM 82: Proceedings of the International Conference on Generalised Linear Models. Edited by R. Gilchrist. V, 188 pages, 1982.

Vol. 15: K.R.W. Brewer and M. Hanif, Sampling with Unequal Probabilities. IX, 164 pages, 1983.

Vol. 16: Specifying Statistical Models: From Parametric to Non-Parametric, Using Bayesian or Non-Bayesian Approaches. Edited by J.P. Florens, M. Mouchart, J.P. Raoult, L. Simar, and A.F.M. Smith, XI, 204 pages, 1983.

Vol. 17: I.V. Basawa and D.J. Scott, Asymptotic Optimal Inference for Non-Ergodic Models. IX, 170 pages, 1983.

Vol. 18: W. Britton, Conjugate Duality and the Exponential Fourier Spectrum. V, 226 pages, 1983.

Vol. 19: L. Fernholz, von Mises Calculus For Statistical Functionals. VIII, 124 pages, 1983.

Vol. 20: Mathematical Learning Models — Theory and Algorithms: Proceedings of a Conference. Edited by U. Herkenrath, D. Kalin, W. Vogel. XIV, 226 pages, 1983.

Vol. 21: H. Tong, Threshold Models in Non-linear Time Series Analysis. X, 323 pages, 1983.

Vol. 22: S. Johansen, Functional Relations, Random Coefficients and Nonlinear Regression with Application to Kinetic Data, VIII, 126 pages, 1984.

Vol. 23: D.G. Saphire, Estimation of Victimization Prevalence Using Data from the National Crime Survey. V, 165 pages, 1984.

Vol. 24: T.S. Rao, M.M. Gabr, An Introduction to Bispectral Analysis and Bilinear Time Series Models. VIII, 280 pages, 1984.

Vol. 25: Time Series Analysis of Irregularly Observed Data. Proceedings, 1983. Edited by E. Parzen. VII, 363 pages, 1984.

Vol. 26: Robust and Nonlinear Time Series Analysis. Proceedings, 1983. Edited by J. Franke, W. Härdle and D. Martin. IX, 286 pages, 1984.

Vol. 27: A. Janssen, H. Milbrodt, H. Strasser, Infinitely Divisible Statistical Experiments. VI, 163 pages, 1985.

Vol. 28: S. Amari, Differential-Geometrical Methods in Statistics. V, 290 pages, 1985.

Vol. 29: Statistics in Ornithology. Edited by B.J.T. Morgan and P.M. North. XXV, 418 pages, 1985.

Vol 30: J. Grandell, Stochastic Models of Air Pollutant Concentration. V, 110 pages, 1985.

Vol. 31: J. Pfanzagl, Asymptotic Expansions for General Statistical Models. VII, 505 pages, 1985.

Vol. 32: Generalized Linear Models. Proceedings, 1985. Edited by R. Gilchrist, B. Francis and J. Whittaker. VI, 178 pages, 1985.

Vol. 33: M. Csörgo, S. Csörgo, L. Horváth, An Asymptotic Theory for Empirical Reliability and Concentration Processes. V, 171 pages, 1986.

Vol. 34: D.E. Critchlow, Metric Methods for Analyzing Partially Ranked Data. X, 216 pages, 1985.

Vol. 35: Linear Statistical Inference. Proceedings, 1984. Edited by T. Calinski and W. Klonecki. VI, 318 pages, 1985.

Vol. 36: B. Matérn, Spatial Variation. Second Edition. 151 pages, 1986.

Vol. 37: Advances in Order Restricted Statistical Inference. Proceedings, 1985. Edited by R. Dykstra, T. Robertson and F.T. Wright. VIII, 295 pages, 1986.

Vol. 38: Survey Research Designs: Towards a Better Understanding of Their Costs and Benefits. Edited by R.W. Pearson and R.F. Boruch. V, 129 pages, 1986.

Vol. 39: J.D. Malley, Optimal Unbiased Estimation of Variance Components. IX, 146 pages, 1986.

Vol. 40: H.R. Lerche, Boundary Crossing of Brownian Motion. V, 142 pages, 1986.

Vol. 41: F. Baccelli, P. Brémaud, Palm Probabilities and Stationary Queues. VII, 106 pages, 1987.

Vol. 42: S. Kullback, J.C. Keegel, J.H. Kullback, Topics in Statistical Information Theory. IX, 158 pages, 1987.

Vol. 43: B.C. Arnold, Majorization and the Lorenz Order: A Brief Introduction. VI, 122 pages, 1987.

Vol. 44: D.L. McLeish, Christopher G. Small, The Theory and Applications of Statistical Inference Functions. VI, 124 pages, 1987.

Vol. 45: J.K. Ghosh (Ed.), Statistical Information and Likelihood. 384 pages, 1988.

Vol. 46: H.-G. Müller, Nonparametric Regression Analysis of Longitudinal Data. VI, 199 pages, 1988.

Vol. 47: A.J. Getson, F.C. Hsuan, {2}-Inverses and Their Statistical Application. VIII, 110 pages, 1988.

Vol. 48: G.L. Bretthorst, Bayesian Spectrum Analysis and Parameter Estimation. XII, 209 pages, 1988.

Vol. 49: S.L. Lauritzen, Extremal Families and Systems of Sufficient Statistics. XV, 268 pages, 1988.

Vol. 50: O.E. Barndorff-Nielsen, Parametric Statistical Models and Likelihood. VII, 276 pages, 1988.

Vol. 51: J. Hüsler, R.-D. Reiss (Eds.), Extreme Value Theory. Proceedings, 1987. X, 279 pages, 1989.

Vol. 52: P.K. Goel, T. Ramalingam, The Matching Methodology: Some Statistical Properties. VIII, 152 pages, 1989.

Vol. 53: B.C. Arnold, N. Balakrishnan, Relations, Bounds and Approximations for Order Statistics. IX, 173 pages, 1989.

Vol. 54: K.R. Shah, B.K. Sinha, Theory of Optimal Designs. VIII, 171 pages, 1989.

Vol. 55: L. McDonald, B. Manly, J. Lockwood, J. Logan (Eds.), Estimation and Analysis of Insect Populations. Proceedings, 1988. XIV, 492 pages, 1989.

Vol. 56: J.K. Lindsey, The Analysis of Categorical Data Using GLIM. V, 168 pages, 1989.

Vol. 57: A. Decarli, B.J. Francis, R. Gilchrist, G.U.H. Seeber (Eds.), Statistical Modelling. Proceedings, 1989. IX, 343 pages, 1989.

Vol. 58: O.E. Barndorff-Nielsen, P. Blæsild, P.S. Eriksen, Decomposition and Invariance of Measures, and Statistical Transformation Models. V, 147 pages, 1989.

Vol. 59: S. Gupta, R. Mukerjee, A Calculus for Factorial Arrangements. VI, 126 pages, 1989.

Vol. 60: L. Györfi, W. Härdle, P. Sarda, Ph. Vieu, Nonparametric Curve Estimation from Time Series. VIII, 153 pages, 1989.

Vol. 61: J. Breckling, The Analysis of Directional Time Series: Applications to Wind Speed and Direction. VIII, 238 pages, 1989.

Vol. 62: J.C. Akkerboom, Testing Problems with Linear or Angular Inequality Constraints. XII, 291 pages, 1990.

Vol. 63: J. Pfanzagl, Estimation in Semiparametric Models: Some Recent Developments. III, 112 pages, 1990.

Vol. 64: S. Gabler, Minimax Solutions in Sampling from Finite Populations. V, 132 pages, 1990.

Vol. 65: A. Janssen, D.M. Mason, Non-Standard Rank Tests. VI, 252 pages, 1990.

Vol. 66: T. Wright, Exact Confidence Bounds when Sampling from Small Finite Universes. XVI, 431 pages, 1991.

Vol. 67: M.A. Tanner, Tools for Statistical Inference: Observed Data and Data Augmentation Methods. VI, 110 pages, 1991.

Vol. 68: M. Taniguchi, Higher Order Asymptotic Theory for Time Series Analysis. VIII, 160 pages, 1991.

Vol. 69: N.J.D. Nagelkerke, Maximum Likelihood Estimation of Functional Relationships. V, 110 pages, 1992.

Vol. 70: K. Iida, Studies on the Optimal Search Plan. VIII, 130 pages, 1992.

Vol. 71: E.M.R.A. Engel, A Road to Randomness in Physical Systems. IX, 155 pages, 1992.

Vol. 72: J.K. Lindsey, The Analysis of Stochastic Processes using GLIM. VI, 294 pages, 1992.

Vol. 73: B.C. Arnold, E. Castillo, J.-M. Sarabia, Conditionally Specified Distributions. XIII, 151 pages, 1992.

Vol. 74: P. Barone, A. Frigessi, M. Piccioni, Stochastic Models, Statistical Methods, and Algorithms in Image Analysis. VI, 258 pages, 1992.

Vol. 75: P.K. Goel, N.S. Iyengar (Eds.), Bayesian Analysis in Statistics and Econometrics. XI, 410 pages, 1992.

Vol. 76: L. Bondesson, Generalized Gamma Convolutions and Related Classes of Distributions and Densities. VIII, 173 pages, 1992.